INTELLIGENT CONTROL
Fuzzy Logic Applications

Clarence W. de Silva

CRC Press
Boca Raton   New York   London   Tokyo

Library of Congress Cataloging-in-Publication Data

de Silva, Clarence W.
   Intelligent control: fuzzy logic applications / Clarence W. de Silva.
      p.   cm.
   Includes bibliographical references and index.
   ISBN 0-8493-7982-2 (acid-free paper)
   1. Intelligent control systems. 2. Fuzzy systems. I. Title.
TJ217.5.D47   1995
629.8—dc20   94-44558
                                                                          CIP

This book contains information obtained from authentic and highly regarded sources. Reprinted material is quoted with permission, and sources are indicated. A wide variety of references are listed. Reasonable efforts have been made to publish reliable data and information, but the author and the publisher cannot assume responsibility for the validity of all materials or for the consequences of their use.

Neither this book nor any part may be reproduced or transmitted in any form or by any means, electronic or mechanical, including photocopying, microfilming, and recording, or by any information storage or retrieval system, without prior permission in writing from the publisher.

CRC Press, Inc.'s consent does not extend to copying for general distribution, for promotion, for creating new works, or for resale. Specific permission must be obtained in writing from CRC Press for such copying.

Direct all inquiries to CRC Press, Inc., 2000 Corporate Blvd., N.W., Boca Raton, Florida 33431.

© 1995 by CRC Press, Inc.

No claim to original U.S. Government works
International Standard Book Number 0-8493-7982-2
Library of Congress Card Number 94-44558
Printed in the United States of America 1 2 3 4 5 6 7 8 9 0
Printed on acid-free paper

Mechatronics Series

Series Editor
S. Shankar Sastry
University of California, Berkeley

Titles Included in the Series

Flexible Joint Robots
Mark C. Readman

Intelligent Control: Fuzzy Logic Applications
Clarence W. de Silva

A Mathematical Introduction to Robotic Manipulation
Richard M. Murray, Zexiang Li, and S. Shankar Sastry

PREFACE

This book is an outgrowth of research, teaching, and industrial applications in the field of Intelligent and Fuzzy Logic Control, which I have been involved in during the past 10 years. The notes that evolved into the manuscript were first developed for several short courses that I taught in Singapore and Vancouver. Of necessity, the mathematical complexities were minimized in the preliminary versions of the manuscript, but were subsequently integrated into the book, for the benefit of advanced students and researchers.

The book is suitable for an undergraduate or an introductory-level graduate course on the subject, and also as a convenient reference tool for practicing professionals. No prior knowledge of the subject is assumed, and much of the book may be comprehended without an advanced knowledge in mathematics. Applications are emphasized throughout the book, and examples and case studies are presented to maintain this flavor. Also, software for developing intelligent control applications are discussed, giving practical details. Advanced mathematical concepts have been introduced without hampering the readability of the book for the novice.

The book consists of 10 chapters and 2 appendices. First, the traditional control techniques are introduced and contrasted with intelligent control. Several methods of representing and processing knowledge are presented, as this is the basis of intelligent control. Subsequently, fuzzy logic is introduced as one such method for representing and processing knowledge. Here, advantages of fuzzy logic over other techniques are pointed out, while indicating some limitations as well. In particular, a hierarchical control structure that is appropriate for use in intelligent control systems is described and further developed. Several applications are introduced. Most of them fall into the areas of Robotics and Mechatronics. However, other applications such as air-conditioning and process/production control are also presented in the book. Some advanced analytical concepts of fuzzy logic are given in Appendix A. Appendix B describes a commercially available software system for developing fuzzy logic applications.

The book includes worked examples, exercises, problems, solutions or answers to selected problems, and references for further reading. Hints are given for research and development avenues of the subject which the reader could pursue while using the book. In summary, the main features of the book include the use of graphic representations and examples to present advanced theory, while removing advanced mathematics and the associated jargon into appendices and research oriented sections; the emphasis on practical applications, giving examples and case studies; and the description of commercial tools for developing applications of fuzzy logic control.

Clarence W. de Silva
Vancouver, Canada

THE AUTHOR

Clarence W. de Silva, Ph.D., is the Natural Sciences and Engineering Research Council (NSERC) Senior Chair Professor of Industrial Automation in the Department of Mechanical Engineering, University of British Columbia, Vancouver, Canada. He obtained his Ph.D. in Dynamic Systems and Control from the Massachusetts Institute of Technology.

Dr. de Silva has authored or co-authored 11 books; about 100 journal papers; and numerous conference papers, bound volumes, and reports. He has been a technical consultant to several leading companies, including IBM and Westinghouse. He also serves on the editorial boards of six international journals.

Dr. de Silva is a Fellow of the American Society of Mechanical Engineers (ASME). He has also been a Fellow of Lilly Foundation, the National Aeronautics and Space Agency/American Society of Engineering Education (NASA/ASEE), Fulbright Program, Advanced Systems Institute of British Columbia, and Killam Program. His research areas include intelligent control, robotics, sensors, actuators, instrumentation, and industrial automation.

ACKNOWLEDGMENTS

This book is an outgrowth of many years of research and teaching. As such, credit must be given to many people who have supported and assisted these activities. While it is impossible to list here all individuals and organizations who have contributed to the development of this book, I wish to name several, albeit at the risk of omitting some important names.

Financial assistance for my research activities has been provided by many organizations, particularly, Natural Sciences and Engineering Research Council of Canada (NSERC), Advanced Systems Institute of British Columbia, B.C. Packers, Ltd., Garfield Weston Foundation, B.C. Hydro and Power Authority, B.C. Ministry of Environment, Network of Centres of Excellence (IRIS), and The University of British Columbia (UBC). Killam Memorial Faculty Fellowships Program deserves a special mention in this regard.

I have been fortunate to serve under two excellent department heads and a dean, Professor Martha Salcudean, Professor Bob Evans, and Professor Axel Meisen. I am grateful to them for providing a conducive environment at UBC for research and teaching, results of which are amply evident in this book. Also, I am grateful to Professor Jim A. N. Poo, Director, Postgraduate School of Engineering, National University of Singapore, who has supported and collaborated in my professional activities. My students and research staff in the Industrial Automation Laboratory at UBC have provided research assistance for my work, some of which is reported in this book. Major contributors in this regard are Dr. Ming Wu, Dr. Nalin Wickramarachchi, Mr. Scott J. Gu, and Mr. Simshon Barlev. The first draft of the manuscript was skillfully typed by Mrs. Edna Dahamarane.

Much of the book represents the results of original research that has been carried out by me with the assistance of my co-workers. These results have been published in journals and conference proceedings, as cited throughout the book. Specifically, I wish to thank The Institute of Electrical and Electronics Engineers, Inc. (IEEE), Elsevier Science, Togai Infralogic, Inc., John Wiley & Sons, and Measurements and Data Corporation for granting permission to include such material in this book. Even though it is difficult to indicate the source for each piece of information, because what is included in the book are not exact reproductions, some correspondence may be pointed out. Notably, the following portions of the work are based on the following sources:

1. Chapter 5 (sections 2, 3, and 4): de Silva, C.W. (1991). An Analytical Framework for Knowledge-Based Tuning, *Engin. Appl. Artif. Intel.*, Vol. 4, No. 3, pp. 177–189.
2. Chapter 6 (sections 2 and 3): de Silva, C.W. (1995). Applications of Fuzzy Logic in the Control of Robotic Manipulators, *Fuzzy Sets and Systems,* Vol. 70, No. 2–3, pp. 223–234.
3. Chapter 7 (sections 4 and 5): de Silva, C.W. (1994). A Criterion for Knowledge Base Decoupling in Fuzzy Logic Control Systems, *IEEE Transactions on Systems, Man, and Cybernetics*, Vol. 24, No. 10, pp. 1548–1552.
4. Chapter 8: Chapter on "Considerations of Hierarchical Fuzzy Control" by C.W. de Silva, in the book, *Theoretical Aspects of Fuzzy Control*, Copyright 1995, John Wiley & Sons, New York. All rights reserved; de Silva, C.W. (1991). Fuzzy Information and Degree of Resolution Within the Context of a Control Hierarchy, *Proc. IEEE Int. Conf. Indust. Electronics, Control, and Instrumentation*, Kobe, Japan, pp. 1590–1595; de Silva, C.W. (1993). Hierarchical Processing of Information in Fuzzy Logic Control Applications, *Proc. 2nd IEEE Conf. Control Applications*, Vancouver, Canada, Vol. 7, No. 5, pp. 457–461.
5. Chapter 9: de Silva, C.W. (1993). Soft Automation of Industrial Processes, *Engin. Appl. Artif. Intel.*, Vol. 6, No. 2, pp. 87–90; de Silva, C.W. and Gu, J. (1994). An Intelligent System for Dynamic Sharing of Workcell Components in Process Automation, *Engin. Appl. Artif. Intel.*, Vol. 7, No. 5, pp. 471–477.
6. Chapter 10: de Silva, C.W. (1994). Automation Intelligence, *Engin. Appl. Artif. Intel.*, Vol. 7, No. 5, pp. 471–477.

I wish to thank the staff of CRC Press, particularly the editors, Ms. Felicia Shapiro and Ms. Julie Haydu, for their excellent contributions in production of the book. Finally, my wife and children deserve a special mention here for their understanding and patience while my time was devoted to research, teaching, supervision, travel, meetings, and book writing!

CONTENTS

1. Conventional and Intelligent Control 1
   Introduction 1
   Conventional Control Techniques 4
   Summary 19
   Problems 19
   References 22

2. Knowledge Representation and Processing 23
   Introduction 23
   Knowledge and Intelligence 24
   Logic 25
   Semantic Networks 30
   Frames 31
   Production Systems 34
   Summary 38
   Problems 39
   References 42

3. Fundamentals of Fuzzy Logic 43
   Introduction 43
   Fuzzy Sets 44
   Fuzzy Logic Operations 47
   Some Definitions 48
   Fuzzy Relations 50
   Composition and Inference 55
   Membership Function Estimation 60
   Summary 62
   Problems 63
   References 68

4. Fuzzy Logic Control 69
   Introduction 69
   Basics of Fuzzy Control 71
   Decision Making with Crisp Measurements 75
   Defuzzification 78
   Architectures of Fuzzy Control 81
   Summary 90
   Problems 92
   References 101

5. Knowledge-Based Tuning 103
   Introduction 103
   Theoretical Background 105

Analytical Framework	111
Computational Efficiency	117
Dynamic Switching of Fuzzy Resolution	122
Illustrative Example	131
Summary	137
Problems	140
References	143

6. Knowledge-Based Control of Robots — 145
- Introduction — 145
- Robotic Control System — 148
- Application to Robots — 149
- In-Loop Direct Control — 152
- High-Level Fuzzy Control — 156
- Control Hierarchy — 160
- System Development — 165
- Servo Expert Development — 170
- Summary — 202
- Problems — 202
- References — 206

7. Servo Motor Tuning — 211
- Introduction — 211
- System Development — 213
- Results — 221
- Theory of Rule Base Decoupling — 224
- Experimental Illustration — 228
- Summary — 231
- Problems — 232
- References — 234

8. Hierarchical Fuzzy Control — 235
- Introduction — 235
- General Concepts — 237
- Hierarchical Model — 241
- Effect of Information Processing — 249
- Application in Process Control — 252
- Summary — 262
- Problems — 264
- References — 267

9. Intelligent Restructuring of Production Systems — 269
- Introduction — 269
- Theoretical Framework — 270
- Implementation Using a Blackboard Architecture — 281
- Case Study — 290

Summary 295
Problems 297
References 299

10. Future Applications 301
Introduction 301
Intelligence in Automation 301
Intelligent Multiagent Control 303
Reconfigurable Autonomous Manipulators 304
Intelligent Fusion of Sensors and Actuators 305
Mechatronics Era 308
Conclusion 310
Problems 311
References 312

Appendix A: Further Topics on Fuzzy Logic 313

Appendix B: Software Tools for Fuzzy Logic Applications 327

Index 337

This book is dedicated to my mother and to the memory of my father.

"Intelligent action cannot be maximized without some guiding principles or theory..."

Joseph Getzels
1956

1 CONVENTIONAL AND INTELLIGENT CONTROL

INTRODUCTION

Fuzzy logic control belongs to the class of "intelligent control", "knowledge-based control", or "expert control". Even though it is incorrect to claim that fuzzy logic control is always better than the conventional types of "crisp" ("hard" or "inflexible") control, it has several advantages. Our objective in this chapter is to outline several of the prominent techniques of control, both modern and classical, and indicate drawbacks of these methods which make "soft" intelligent control more appropriate in some applications.

With steady advances in the field of *artificial intelligence (AI)*, especially pertaining to the development of practical *expert systems or knowledge systems*, there has been considerable interest in using AI for controlling complex processes (Francis and Leitch, 1984; Goff, 1985; Isik and Mystel, 1986; Saridis, 1988; de Silva, 1991). The rationale for using intelligent control systems may be related to difficulties which are commonly experienced by practicing control engineers. It is generally difficult to accurately represent a complex process by a mathematical model or by a simple computer model. Even when the model itself is tractable, control of the process using a hard control algorithm might not provide satisfactory performance. Furthermore, it is commonly known that the performance of some industrial processes can be considerably improved through the higher-level control actions (tuning actions, in particular) made by an experienced and skilled operator, and these actions normally cannot be formulated by hard control algorithms.

There are many reasons for the practical deficiencies of hard (crisp) control algorithms. If the process model itself is inaccurate, a conventional, *model-based control* can generate unsatisfactory results. Even with an accurate model, if the parameter values are partially known, ambiguous, or vague, appropriate estimates must be made. Crisp control algorithms that are based on such *incomplete information* will not usually give satisfactory results. The environment with which the process interacts may not be completely predictable, and it is normally not possible for a hard algorithm to accurately respond to a condition that it did not anticipate and that it could not "understand". A hard control algorithm will need a complete set of data to produce results, and indeed, program instructions and data are conventionally combined into the same memory of the controller. If some data were not available, say, unexpectedly as a result of sensing problems, the inflexible algorithm will naturally fail.

Humans are flexible (or soft) systems. They can adapt to unfamiliar situations, and they are able to gather information in an efficient manner and discard irrelevant details. The information which is gathered need not be complete and precise and could be *general, qualitative*, and *vague* because humans can reason, infer, and deduce new information and knowledge. They have common sense. They can make good decisions, and also can provide logical explanations for those decisions. They can *learn, perceive*, and improve their skills through experience. Humans can be creative, inventive, and innovative. It is a very challenging task to seek to develop systems that possess even a few of these simple human abilities. This is the challenge faced by workers in the field of AI and *knowledge-based systems*. However, humans have weaknesses too. For example, they tend to be slow, inaccurate, forgetful, and emotional. In other areas, of course, one might argue that these human qualities are strengths rather than weaknesses.

In a knowledge-based system one seeks to combine the advantages of a computer with some of the intelligent characteristics of a human for making *inferences* and decisions. It is clearly not advisable to attempt simply to mimic the human thought process without carefully studying the needs of the problem at hand. In fact, a faithful duplication is hardly feasible. Rather, in the initial stages of development of this field it is adequate to develop special-purpose systems that cater to a limited class of problems. Knowledge-based control or intelligent control is one such development.

Example

An "iron butcher" is a head-cutting machine that is commonly used in the fish processing industry. Millions of dollars worth of salmon are wasted annually due to inaccurate head cutting using these somewhat

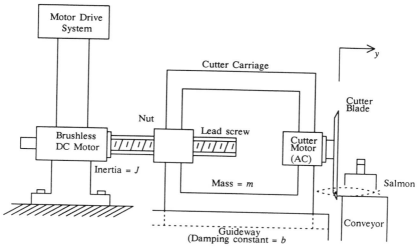

Figure 1 A positioning system for an automated fish cutting machine.

outdated machines. The main cause of wastage is the "over-feed problem". This occurs when a salmon is inaccurately positioned with respect to the cutter blade so that the cutting location is beyond the collar bone and into the body of a salmon. An effort has been made to correct this situation by sensing the position of the collar bone using a vision system and automatically positioning the cutter blade accordingly. A schematic representation of an electromechanical positioning system of a salmon head cutter is shown in Figure 1. Positioning of the cutter is achieved through a lead screw and nut arrangement which is driven by a brushless DC motor. The cutter carriage is integral with the nut of the lead screw and the AC motor that drives the cutter blade. The carriage slides along a lubricated guideway. In this system position control of the cutter and the speed control of the conveyor can be accomplished using conventional means (e.g., through servo control). However, suppose that in addition the quality of the processed fish is to be observed and on that basis several control adjustments are to be made. These adjustments may include descriptive (e.g., small, medium, large) changes in the cutter blade speed and conveyor speed. The required degree of adjustment is learned through experience. It is not generally possible to establish complex control algorithms for such adjustments, and the control knowledge may be expressed as a set of rules such as:

> If the "quality" of the processed fish is "not acceptable", and if the cutting load appears to be "high", then "moderately" decrease the conveyor speed, or else if the hold-down force appears to be "low", then "slightly" increase the holding force or else...

4 INTELLIGENT CONTROL: FUZZY LOGIC APPLICATIONS

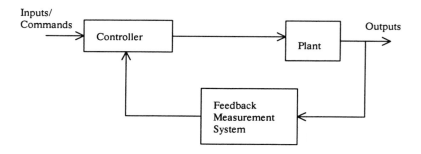

Figure 2 Schematic representation of a control system.

Such a set of rules forms the *knowledge base* of this control problem. It should be clear that this rule base has qualitative, descriptive, and linguistic terms such as "quality", "acceptable", "high", and "slight". Crisp representations and algorithms used in conventional control cannot directly incorporate such knowledge.

■ CONVENTIONAL CONTROL TECHNIQUES

Before we study the subject of intelligent control let us briefly review some of the more predominant techniques used in traditional control systems. The main components of a control system are

1. The *plant*, or the process that is being controlled
2. The *controller*, which controls the plant
3. The *measurement system,* which is needed for *feedback control*

A schematic diagram of this interpretation is shown in Figure 2.

By control, we mean making the plant respond to inputs in a desired manner. In a *regulator*-type control system the objective is to maintain the output at a desired (constant) value. In a *servomechanism*-type control system the objective is for the output to follow a desired trajectory. In the design of a control system the desired behavior in response is usually guaranteed by meeting a set of *specifications*, usually expressed with respect to the attributes given in Table 1. These desired attributes are further discussed in standard books on control (see, for example, de Silva and Aronson, 1986; de Silva, 1989).

Note in Figure 2 that if the plant is stable and is completely and accurately known, and if the inputs to the plant from the controller can be generated and applied accurately, accurate control will be possible even without feedback control. In this case the measurement system is not needed (or at least not needed for feedback) and thus we have an *open-loop control* system.

TABLE 1
PERFORMANCE SPECIFICATION FOR A CONTROL SYSTEM

Attribute	Desired value	Purpose	Specifications
Stability	High	The response does not grow without limit and decays to the desired value	Percentage overshoot, settling time, pole (eigenvalue) locations, time constants, phase and gain margins, damping ratios
Speed of response	Fast	The plant responds quickly to inputs	Rise time, peak time, delay time, natural frequencies, resonant frequencies, bandwidth
Steady-state error	Low	The offset from the desired response is negligible	Error tolerance for a step input
Robustness	High	Accurate response under uncertain conditions (signal noise, model error, etc.) and under parameter variation	Input noise tolerance, measurement error tolerance, model error tolerance
Dynamic interaction	Low	One input affects only one output	Cross-sensitivity, cross-transfer functions

The controller and the measurement system both may contain a *signal preprocessor* or a *prefilter*. The purpose of a prefilter is to shape a signal, which will generally result in improved performance (for example, through noise rejection or by producing a signal that is more compatible with the component to which it is fed). Similarly, a special *compensator* may be added to the controller. Note that as long as a feedback loop exists, the controller, regardless of its location within the loop, is called a feedback controller.

When designing a control system, the engineer is expected to establish controllers and compensators, including control algorithms and hardware, that will make the plant behave in a required manner. The task is not simple and will depend on many factors such as nature of the plant (e.g., size, complexity, nonlinearity, time-varying parameters, dynamic interactions), control objective and specifications, and cost considerations. In particular, it is necessary to select a control technique. Some of the more common techniques are outlined below.

Proportional-Integral-Derivative Servo Control

A block diagram for a system with proportional-integral-derivative (PID) control is shown in Figure 3. The controller parameters $K_p, \tau_i,$ and τ_d are defined in Table 2 and their primary functions and side effects are listed. Although a single-input-single-output (SISO) system is represented in Figure 3, the idea can be extended to the

6 INTELLIGENT CONTROL: FUZZY LOGIC APPLICATIONS

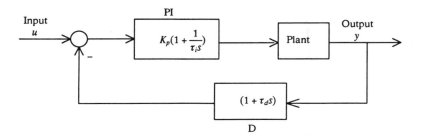

Figure 3 A typical PID servo controller.

TABLE 2
CHARACTERISTICS OF PID CONTROL

Controller parameter	Description	Functions	Undesirable side effects
K_p	Proportional gain	Speeds up the response Reduces offset Reduces cross-coupling	System can become less stable (overshoot, oscillations, etc.)
τ_i	Integral time constant	Reduces offset Reduces noise	Can slow down the system Has a destabilizing effect Introduces a phase lag
τ_d	Derivative time constant	Stabilizes the response (damping) Speeds up the system Provides a phase lead (anticipatory effect)	Enhances high-frequency noise Difficult to physically implement

multivariable or multi-input-multi-output (MIMO) case. Then u and y become vectors. Also, in practice compensating circuitry (e.g., lead-lag) are also added to a servo system of the type shown in Figure 3.

The problem of controller design, in the present case, is a matter of selecting the values for the PID parameters K_p, τ_i, and τ_d. If the plant is assumed linear, and if a reasonably accurate model is available for it, the parameter selection problem is relatively simple. For example, tables, charts, and computer programs based on both time-domain techniques (e.g., to satisfy specifications for rise time, percentage overshoot, and steady-state error) and frequency-domain techniques (e.g., satisfying specifications on gain and phase margins, bandwidth, and pole locations using Bode, Nyquist, and root locus tools) are commonly available. When the plant is unknown, the PID design problem is essentially one of trial and error, and it becomes increasingly difficult in the presence of large dynamic interactions, nonlinearities and parameter variations in the plant. In particular, the PID parameters must be continuously adjusted (or "tuned") in order to meet the performance criteria. Conventional tuning algorithms of this type are "crisp" and assume linear plants. Furthermore, tuning is not "intelligent" in these conventional methods. Clearly, mistakes made in the past and the experience gained in operating the system are not "learned" and made use of later to obtain increasingly better performance.

Nonlinear Feedback Control
Simple, linear servo control is known to be inadequate for transient and high-speed operation of complex plants. Past experience of servo control in process applications is extensive, however, and servo control is extensively used in many commercial applications (e.g., robots). For this type of control to be effective, however, nonlinearities and dynamic coupling must be compensated faster than the control bandwidth at the servo level. One way of accomplishing this is by implementing a linearizing and decoupling controller inside the servo loops. This technique is termed *feedback linearization*. One such technique that is useful in controlling nonlinear and coupled dynamic systems such as robots is outlined here (see also de Silva and MacFarlane, 1989).

Consider a mechanical dynamic system (plant) given by

$$\mathbf{M}(\mathbf{q})\frac{d^2\mathbf{q}}{dt^2} = \mathbf{n}\left(\mathbf{q}, \frac{d\mathbf{q}}{dt}\right) + \mathbf{f}(t) \tag{1}$$

in which

$\mathbf{f} = \begin{bmatrix} f_1 \\ f_2 \\ \vdots \\ f_r \end{bmatrix} =$ vector of input forces at various locations of the system

$\mathbf{q} = \begin{bmatrix} q_1 \\ q_2 \\ \vdots \\ q_r \end{bmatrix} =$ vector of response variables (e.g., positions) at the forcing locations of the system

$\mathbf{M}(\mathbf{q}) =$ inertia matrix (nonlinear), and

$\mathbf{n}\left(\mathbf{q}, \frac{d\mathbf{q}}{dt}\right) =$ a vector of remaining nonlinear effects in the system (e.g., damping, backlash, gravitational effects)

Now suppose that we can model **M** by **M̂** and **n** by **n̂**. Then, let us use the nonlinear (linearizing) feedback controller given by

$$\mathbf{f} = \hat{\mathbf{M}}\mathbf{K}\left[\mathbf{e} + \mathbf{T}_i^{-1}\int \mathbf{e}\,dt - \mathbf{T}_d \frac{d\mathbf{q}}{dt}\right] - \hat{\mathbf{n}} \qquad (2)$$

in which

$$\mathbf{e} = \mathbf{q}_d - \mathbf{q} = \text{error (correction) vector}$$
$$\mathbf{q}_d = \text{desired response}$$

and **K**, $\mathbf{T}_i$, and $\mathbf{T}_d$ are constant control parameter matrices. This control scheme is shown in Figure 4. By substituting the controller equation (2) into the plant equation (1) we get

$$\mathbf{M}\frac{d^2\mathbf{q}}{dt^2} = \mathbf{n} - \hat{\mathbf{n}} + \hat{\mathbf{M}}\mathbf{K}\left[\mathbf{e} + \mathbf{T}_i^{-1}\int \mathbf{e}\,dt - \mathbf{T}_d \frac{d\mathbf{q}}{dt}\right] \qquad (3)$$

If our models are exact, we have **M** = **M̂** and **n** = **n̂**. Then, because the inverse of matrix **M̂** exists in general (because the inertia matrix is *positive definite*), we get

$$\frac{d^2\mathbf{q}}{dt^2} = \mathbf{K}\left[\mathbf{e} + \mathbf{T}_i^{-1}\int \mathbf{e}\,dt - \mathbf{T}_d \frac{d\mathbf{q}}{dt}\right] \qquad (4)$$

Equation (4) represents a linear, constant parameter system with PID control. The proportional control parameters are given by the gain matrix **K**, the integral control parameters by $\mathbf{T}_i$, and the derivative control parameters by $\mathbf{T}_d$. It should be clear that we are free to select these parameters so as to achieve the desired response. In particular, if these three parameter matrices are chosen to be diagonal, then the control system, as given by Equation (4) and shown in Figure 4, will be uncoupled (i.e., one input affects only one output) and will not contain dynamic interactions. In summary, this controller has the advantages of linearizing and decoupling the system; its disadvantages are that accurate models will be needed and that the control algorithm is crisp and unable to handle qualitative or partially known information, learning, etc.

Instead of using analytical modeling, the parameters in **M̂** and **n̂** may be obtained through the measurement of various input-output pairs. This is called *model identification,* and can cause further

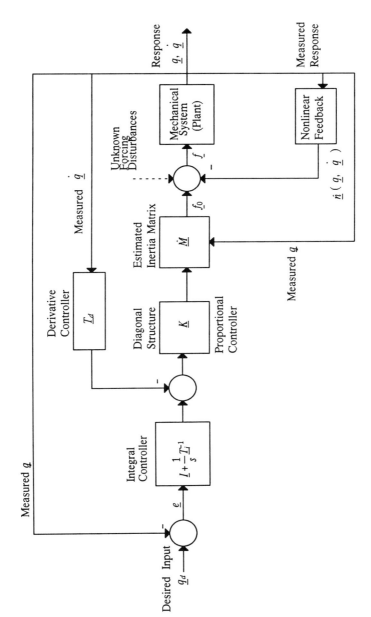

Figure 4 The structure of the model-based nonlinear feedback control system.

complications in terms of instrumentation and data processing speed, particularly because some of the model parameters must be estimated in real time.

Adaptive Control

An adaptive control system is a feedback control system in which the values of some or all of the controller parameters are modified (adapted) during the system operation (in real time) on the basis of some performance measure, when the response (output) requirements are not satisfied. The techniques of adaptive control are numerous because many criteria can be employed for modifying the parameter values of a controller (Zhou and de Silva, 1995). According to the above definition, *self-tuning control* (see, for example, Clarke and Gawthrop, 1985) falls into the same category. In fact, the terms "adaptive control" and "self-tuning control" have been used interchangeably in the technical literature. Performance criteria used in self-tuning control may range from time-response or frequency-response specifications, parameters of "ideal" models, desired locations of poles and zeros, and cost functions. Generally, however, in self-tuning control of a system some form of parameter estimation or identification is performed on-line using input-output measurements from the system, and the controller parameters are modified using these estimated parameter values. A majority of the self-tuning controllers developed in the literature is based on the assumption that the plant (process) is linear and time invariant. This assumption does not generally hold true for complex industrial processes. For this reason we shall restrict our discussion to an adaptive controller that has been developed for nonlinear and coupled plants.

On-line estimation or system identification, which may be required for adaptive control, may be considered to be a preliminary step of "learning". In this context, *learning control* and adaptive control are related, but learning is much more complex and sophisticated than a quantitative estimation of parameter values. In a learning system control decisions are made using the cumulative experience and knowledge gained over a period of time. Furthermore, the definition of learning implies that a learning controller will "remember" and improve its performance with time. This is an evolutionary process that is true for intelligent controllers, but not generally for adaptive controllers.

Here, we briefly describe a *model-referenced adaptive control* (MRAC) technique (also see Dubowsky and Des Forges, 1979; Astrom and Wittenmark, 1989). The general approach of MRAC is illustrated by the block diagram in Figure 5. In nonadaptive feedback control the response measurements are fed back into the drive controller through a feedback controller, but the values of the controller parameters (feedback gains) themselves are unchanged during

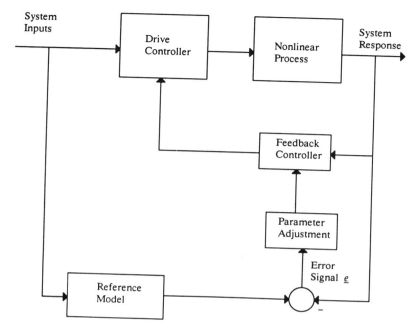

Figure 5 Model-referenced adaptive controller.

operation. In adaptive control these parameter values are changed according to some criterion. In model-referenced adaptive control, in particular, the same reference input that is applied to the physical system is applied to a reference model as well. The difference between the response of the physical system and the output from the reference model is the error. The ideal objective is to make this error zero at all times. Then the system will perform just like the reference model. The error signal is used by the adaptation mechanism to determine the necessary modifications to the values of the controller parameters in order to reach this objective. Note that the reference model is an idealized model which generates a desired response when subjected to the reference input, at least in an asymptotic manner (i.e., the error converges to zero). In this sense it is just a means of performance specification and may not possess any resemblance or analogy to an analytical model of the process itself. For example, the reference model may be chosen as a linear, uncoupled system with desired damping and bandwidth properties (i.e., damping ratios and natural frequencies).

A popular approach to derive the adaptive control algorithm (i.e., the equations expressing how the controller parameters should be changed in real time) is through the use of the MIT rule. In this method, the controller parameters are changed in the direction opposite to the maximum slope of the quadratic error function. Specifically, the quadratic function

$$V(\mathbf{p}) = \mathbf{e}^T \mathbf{W} \mathbf{e} \tag{5}$$

is formed, where **e** is the error signal vector shown in Figure 5, **W** is a diagonal and positive-definite weighting matrix, and **p** is a vector of control parameters which will be changed (adapted) during control. The function V is minimized numerically with respect to **p** during the controller operation, subject to some simplifying assumptions. The details of the algorithm are found in the literature.

The adaptive control algorithm described here has the advantage that it does not necessarily require a model of the plant itself. The reference model can be chosen to specify the required performance, the objective of MRAC being to drive the response of the system toward that of the reference model. Several drawbacks exist in this scheme, however. Because the reference model is quite independent of the plant model, the required control effort could be excessive and the computation itself could be slow. Furthermore, a new control law must be derived for each reference model. Also, the control action must be generated much faster than the speed at which the nonlinear terms of the plant change because the adaptation mechanism has been derived by assuming that some of the nonlinear terms remain more or less constant.

Many other adaptive control schemes depend on a reasonably accurate model of the plant, not just a reference model (de Silva and Van Winssen, 1987). The models may be obtained either analytically or through identification (experimental). Adaptive control has been successfully applied in complex, nonlinear, and coupled systems, even though it has several weaknesses, as mentioned previously.

Sliding Mode Control

Sliding mode control, variable structure control, and suction control fall within the same class of control techniques, and are somewhat synonymous. The control law in this class is generally a switching controller. A variety of switching criteria may be employed. Sliding mode control may be treated as an adaptive control technique. Because the switching surface is not fixed, its variability is somewhat analogous to an adaptation criterion. Specifically, the error of the plant response is zero when the control falls on the sliding surface. Several workers have developed variable-structure control strategies for various applications (e.g., Young, 1978; Slotine, 1985).

Consider a plant that is modeled by the nth order nonlinear ordinary differential equation

$$\frac{d^n y}{dt^n} = f(\mathbf{y},t) + u(t) + d(t) \qquad (6)$$

where
- y = response of interest
- $u(t)$ = input variable
- $d(t)$ = unknown disturbance input
- $f(\bullet)$ = an unknown nonlinear model of the process which depends on the response vector

$$\mathbf{y} = \left[y, \dot{y}, \ldots, \frac{d^{n-1} y}{dt^{n-1}} \right]^T$$

A time-varying sliding surface is defined by the differential equation

$$s(\mathbf{y},t) = \left(\frac{d}{dt} + \lambda \right)^{n-1} \tilde{y} = 0 \qquad (7)$$

with $\lambda > 0$. Note the response error $\tilde{y} = y - y_d$, where y_d is the desired response. Similarly, $\tilde{\mathbf{y}} = \mathbf{y} - \mathbf{y}_d$ may be defined. It should be clear from Equation (7) that if we start with zero initial error ($\tilde{y}(0) = 0$) then $s = 0$ corresponds to $\tilde{y}(t) = 0$ for all t. This will guarantee that the desired trajectory $y_d(t)$ is tracked accurately at all times. Hence, the control objective would be to force the error state vector $\tilde{\mathbf{y}}$ onto the sliding surface $s = 0$. This control objective will be achieved if the control law satisfies

$$\dot{s}\, sgn(s) \le -\eta \quad \text{with} \quad \eta > 0 \qquad (8)$$

where $sgn(s)$ is the *signum function*.

The nonlinear process $f(\mathbf{y},t)$ is generally unknown. Suppose that $f(\mathbf{y},t) = \hat{f}(\mathbf{y},t) + \Delta f(\mathbf{y},t)$, where $\hat{f}(\mathbf{y},t)$ is a completely known function, and Δf represents modeling uncertainty. Specifically, consider the control equation,

$$u = -\hat{f}(\mathbf{y},t) - \sum_{p=1}^{n-1} \binom{n-1}{p} \lambda^p \frac{d^{n-p} \tilde{y}}{dt^{n-p}} - K(\mathbf{y},t) sgn(s) \qquad (9)$$

where $K(\mathbf{y},t)$ is an upper bound for the total uncertainty in the system (i.e., disturbance, model error, speed of error reduction, etc.) and

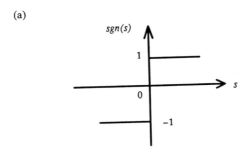

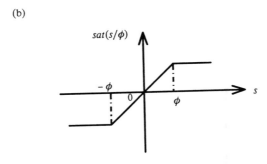

Figure 6 Switching functions used in sliding mode control: (a) signum function, (b) saturation function.

$$\binom{n-1}{p} = \frac{(n-1)!}{p!(n-1-p)!}$$

This sliding-mode controller satisfies Equation (8), but has drawbacks arising from the *sgn(s)* function. Specifically, very high switching frequencies can result when the control effort is significant. This is usually the case in the presence of large modeling errors and disturbances. High-frequency switching control can lead to the excitation of high-frequency modes in the plant. It can also lead to chattering problems. This problem can be reduced if the signum function in Equation (9) is replaced by a *saturation function,* with a boundary layer $\pm\phi$, as shown in Figure 6. In this manner, any switching that would have occurred within the boundary layer would be filtered out. Furthermore, the switching transitions would be much less severe. Clearly, the advantages of sliding mode control include robustness against factors such as nonlinearity, model uncertainties, disturbances, and parameter variations.

Linear Quadratic Gaussian (LQG) Control

This is an *optimal control* technique that is intended for quite linear systems with random input disturbances and output (measurement) noise. Consider the *linear* system given by the set of first order differential equations (*state equations*)

$$\frac{d\mathbf{x}}{dt} = \mathbf{Ax} + \mathbf{Bu} + \mathbf{Fv} \qquad (10)$$

and the output equations

$$\mathbf{y} = \mathbf{Cx} + \mathbf{w} \qquad (11)$$

in which

$$\mathbf{x} = \begin{bmatrix} x_1 \\ x_2 \\ \vdots \\ x_n \end{bmatrix} \text{ is the state vector,}$$

$$\mathbf{u} = \begin{bmatrix} u_1 \\ u_2 \\ \vdots \\ u_r \end{bmatrix} \text{ is the vector of system inputs, and}$$

$$\mathbf{y} = \begin{bmatrix} y_1 \\ y_2 \\ \vdots \\ y_m \end{bmatrix} \text{ is the vector of system outputs.}$$

The vectors **v** and **w** represent input disturbances and output noise, respectively, which are assumed to be white noise (i.e., zero-mean random signals whose *power spectral density function* is flat) with covariance matrices **V** and **W**. Also, **A** is called the system matrix, **B** the *input distribution matrix,* and **C** the *output formation matrix*. In LQG control the objective is to minimize the *performance index* (cost function)

$$J = E\left\{ \int_0^\infty (\mathbf{x}^T \mathbf{Q} \mathbf{x} + \mathbf{u}^T \mathbf{R} \mathbf{u}) dt \right\} \qquad (12)$$

in which **Q** and **R** are diagonal matrices of weighting and E denotes the "expected value" (or mean value) of a random process (Maciejowski, 1989).

In the LQG method the controller is implemented as the two-step process:

1. Obtain the estimate $\hat{\mathbf{x}}$ for the state vector **x** using a *Kalman filter* (with gain $\mathbf{K}_f$).
2. Obtain the control signal as a product of $\hat{\mathbf{x}}$ and a *gain matrix* $\mathbf{K}_0$ by solving a noise-free linear quadratic optimal control problem.

This implementation is shown by the block diagram in Figure 7. It can be analytically shown that the noise-free quadratic optimal controller is given by the gain matrix

$$\mathbf{K}_0 = \mathbf{R}^{-1}\mathbf{B}^T\mathbf{P}_0 \tag{13}$$

where $\mathbf{P}_0$ is the positive semidefinite solution of the *algebraic Riccati equation*

$$\mathbf{A}^T\mathbf{P}_0 + \mathbf{P}_0\mathbf{A} - \mathbf{P}_0\mathbf{B}\mathbf{R}^{-1}\mathbf{B}^T\mathbf{P}_0 + \mathbf{Q} = 0 \tag{14}$$

The Kalman filter is given by the gain matrix

$$\mathbf{K}_f = \mathbf{P}_f\mathbf{C}^T\mathbf{W}^{-1} \tag{15}$$

where $\mathbf{P}_f$ is obtained as before by solving

$$\mathbf{A}\mathbf{P}_f + \mathbf{P}_f\mathbf{A}^T - \mathbf{P}_f\mathbf{C}^T\mathbf{W}^{-1}\mathbf{C}\mathbf{P}_f + \mathbf{F}\mathbf{V}\mathbf{F}^T = 0 \tag{16}$$

An advantage of this controller is that the stability of the closed-loop control system is guaranteed as long as both the plant model and the Kalman filter are *stabilizable* and *detectable*. Note that if uncontrollable states of a system are stable, the system is stabilizable. Similarly, if the unobservable states of a system are stable, the system is detectable. Another advantage is the precision of the controller as long as the underlying assumptions are satisfied, but LQG control is also a model-based "crisp" scheme. Model errors and noise characteristics can significantly affect the performance. Also, even though stability is guaranteed, good stability margins and adequate robustness are not guaranteed in this method. Computational complexity (solution of two Riccati equations) is another drawback.

H_∞ (*H*-Infinity) Control

This is a relatively new optimal control approach which is quite different from the LQG method. However, this method also assumes a

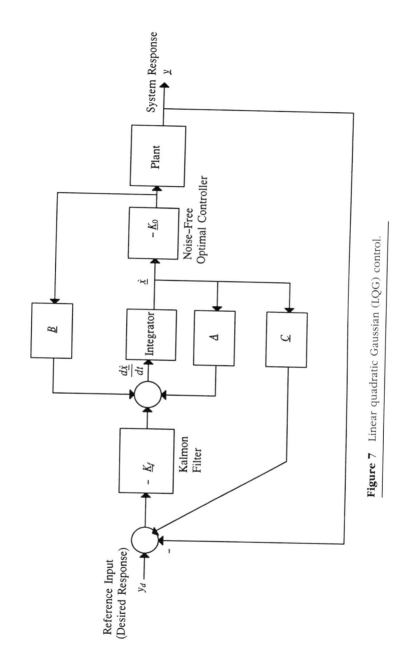

Figure 7 Linear quadratic Gaussian (LQG) control.

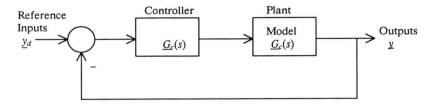

Figure 8 A linear multivariable feedback control system.

linear plant with constant parameters, which may be modeled by a *transfer function* in the SISO case or by a *transfer matrix* in the MIMO case. Without going into the analytical details, let us outline the principle behind H_∞ control (Glover and Doyle, 1988).

Consider the MIMO, linear, feedback control system shown by the block diagram in Figure 8, where **I** is an *identity matrix*. It satisfies the relation

$$\mathbf{GG}_c[\mathbf{y}_d - \mathbf{y}] = \mathbf{y}$$

or

$$[\mathbf{I} + \mathbf{GG}_c]\mathbf{y} = \mathbf{GG}_c\mathbf{y}_d \qquad (17)$$

Because the plant **G** is fixed, the design problem here is to select a suitable controller $\mathbf{G}_c$ that will result in a required performance of the system. In other words, the closed-loop transfer matrix

$$\mathbf{H} = [\mathbf{I} + \mathbf{GG}_c]^{-1}\mathbf{GG}_c \qquad (18)$$

must be properly "shaped" through an appropriate choice of $\mathbf{G}_c$. The required shape of $\mathbf{H}(s)$ may be consistent with the classical specifications such as

1. Unity $|\mathbf{H}(j\omega)|$ or large $|\mathbf{GG}_c(j\omega)|$ at low frequencies in order to obtain small steady-state error for step inputs
2. Small $|\mathbf{H}(j\omega)|$ or small $|\mathbf{GG}_c(j\omega)|$ at large frequencies so that high-frequency noise would not be amplified, and further, the controller would be physically realizable
3. Adequately high gain and phase margins in order to achieve the required stability levels

Of course, in theory there is an "infinite" number of possible choices for $\mathbf{G}_c(s)$ that will satisfy such specifications. The H_∞ method uses an optimal criterion to select one of these "infinite" choices. Specifically, the choice that minimizes the so-called "H_∞ norm" of the closed-loop transfer matrix $\mathbf{H}(s)$, is chosen. The rationale is that this optimal solution is known to provide many desired characteristics (with respect to stability, robustness in the presence of model uncertainty and noise, sensitivity, etc.) in the control system.

The H_∞ norm of a transfer matrix **H** is the maximum value of the largest *singular value* of $\mathbf{H}(j\omega)$, maximum being determined over the entire frequency range. A singular value of $\mathbf{H}(j\omega)$ is the square root of an eigenvalue of the matrix $\mathbf{H}(j\omega)\mathbf{H}^T(j\omega)$.

The H_∞ control method has the advantages of stability and robustness. The disadvantages are that it is a "crisp" control method that is limited to linear, time-invariant systems, and that it is a model-based technique.

■ SUMMARY

Conventional control techniques such as PID control, nonlinear feedback control, adaptive control, sliding mode control, LQG control, and H_∞ control have many advantages. When the values of the controller parameters are known, the control signals are generated exactly. Also, when the underlying assumptions are satisfied, many of these methods provide good stability, robustness to model uncertainties and disturbances, and speed of response. However, there are several disadvantages. The control algorithms are "hard" or "inflexible" and cannot generally handle "soft" intelligent control which may involve reasoning and inference making using incomplete, vague, noncrisp, and qualitative information, and learning and self-organization through past experience and knowledge. Knowledge-based control, expert control, and intelligent control are somewhat synonymous, and fuzzy control is a particular type of intelligent control. The main feature of such control is that a control knowledge base (typically in the form of a set of rules) is available within the controller, and the control actions are generated by applying existing conditions or data to the knowledge base, making use of an inference mechanism. Also, the knowledge base and the inference mechanism can handle noncrisp and incomplete information, and the knowledge itself will improve and evolve through "learning" and past experience.

■ PROBLEMS

1. Specifications used in the design of a control system may include the following:
   (a) Rise time
   (b) Peak time
   (c) Settling time
   (d) Delay time
   (e) Damping ratio
   (f) Undamped natural frequency

(g) Damped natural frequency (h) Resonant frequency
(i) Phase margin (j) Gain margin
(k) Bandwidth (l) Low-frequency gain
(m) Percentage overshoot (n) Gain at resonance

Indicate which ones of these specifications primarily represent:
(i) System stability
(ii) Speed of response
(iii) Steady-state error

Also, which ones are time-domain specifications and which are frequency-domain specifications?

2. Consider a system given by the transfer function

$$G(s) = \frac{(s^3 + p)}{(s^2 + 2\xi\omega_n s + \omega_n^2)}$$

where p, ξ, and ω_n are system parameters and s is the Laplace variable.
(a) What is the frequency transfer function $G(j\omega)$ of this system?
(b) Is this system physically realizable, and why?

3. In sliding mode control the objective is to push the system response toward a so-called "sliding surface". This guarantees zero error throughout the operation if the starting error is zero, otherwise the error will asymptotically reach zero. Explain why the control condition given by Equation (8) satisfies this objective.

4. Suppose that a nonconventional, "intelligent" controller generates the following control inferences:
(a) Increase the control signal to some value.
(b) You may or may not increase the control signal.
(c) Increase the control signal to value (variable) x.
(d) Increase the control signal to 2.5 within a tolerance of ±5%.
(e) Increase the control signal to 2.5 with a probability of 90%.
(f) Increase the control signal slightly.

Indicate which of these inferences is
(i) Vague
(ii) Ambiguous
(iii) General
(iv) Imprecise
(v) Uncertain
(vi) Fuzzy (i.e., noncrisp)

5. Which control method would you recommend for the following applications:
   (a) Servo control of a single-axis positioning table with a permanent-magnet DC motor.
   (b) Control of a low-bandwidth (slow), yet nonlinear, robot. A linear model and upper bounds for the model error and for input disturbances are available.
   (c) Active control of a vehicle suspension system. The system is almost linear. Random disturbance inputs and measurement noise are assumed to be *white noise,* with known covariances.
   (d) Control of a hydraulic-actuated, nonlinear, wood harvesting machine. A reasonably accurate, nonlinear model of the process is available.
   (e) A milling machine which performs quite nonlinear machining operations. A model of the process is not available, but the desired performance of the system may be represented by a linear reference model. The system is computer controlled and the controller parameters are easily and quickly adjustable.
   (f) The positioning system of a magnetic disk drive in a computer workstation. The process is nearly linear and a model is available, but the control system must be quite stable and robust, and should be offset free.
   (g) Control of a rotary cement kiln, a highly nonlinear process, is a difficult problem. Modeling of the process is known to be quite complex and inexact, and a model is not available. The process is reasonably slow, and many control inputs can be manually adjusted. Past experience of the process operators is properly documented. Usually the drive load, temperatures, oxygen content in the exhaust, and the free lime content at the output are observed, and the burner conditions (fuel, air flow rate, etc.) and the material input are adjusted accordingly.

6. In your opinion what is a control system? Indicate the difference between a feedback control system and an open-loop control system.

7. In digital control, what are some of the important considerations that you will have to address, as a control engineer, for instance, when purchasing, designing, or implementing a digital controller?

8. Compare the use of DC servo motors with that of hydraulic servo actuators in a high-speed material processing machine.

REFERENCES

Astrom, K.J. and Wittenmark, B. (1989). *Adaptive Control,* Addison-Wesley, Reading, MA.

Clarke, D.W. and Gawthrop, P.J. (1975). Self-Tuning Controller, *Proc. Inst. Electr. Eng. U.K.,* Vol. 122, No. 9, pp. 929–934.

de Silva, C.W. (1989). *Control Sensors and Actuators,* Prentice-Hall, Englewood Cliffs, NJ.

de Silva, C.W. (1991). An Analytical Framework for Knowledge-Based Tuning of Servo Controllers, *Engin. Appl. Artif. Intell.,* Vol. 4, No. 3, pp. 177–189.

de Silva, C.W. and Aronson, M.H. (1986). *Process Control,* Measurements and Data Corporation, Pittsburgh, PA.

de Silva, C.W. and Van Winssen, J.C. (1987). Least Squares Adaptive Control for Trajectory Following Robots, *ASME J. Dyn. Syst. Measure. Control,* Vol. 109(2), pp. 104–110.

de Silva, C.W. and MacFarlane, A.G.J. (1989). Knowledge-Based Control Approach for Robotic Manipulators, *Int. J. Control,* Vol. 50, No. 1, pp. 249–273.

Dubowsky, S. and Des Forges, D.T. (1979). The Application of Model-Referenced Adaptive Control to Robotic Manipulators, *ASME J. Dyn. Syst. Measure. Control,* Vol. 101, pp. 193–200.

Francis, J.C. and Leitch, R.R. (1985). ARTIFACT: A Real-Time Shell for Intelligent Feedback Control, *Research and Development in Expert Systems,* Bramer, M.A. (Ed.), Cambridge University Press, Cambridge, U.K.

Glover, K. and Doyle, J.C. (1988). State-Space Formulae for All Stabilizing Controllers that Satisfy an H_∞ Norm Bound and Relations to Risk Sensitivity, *Systems and Control Letters,* Vol. II, pp. 167–172.

Goff, K.W. (1985). Artificial Intelligence in Process Control, *Mech. Eng. ASME,* Vol. 107, No. 10, pp. 53–57.

Isik, C. and Mystel, A. (1986). Decision Making at a Level of Hierarchical Control for Unmanned Robots, *Proc. 1986 IEEE Int. Conf. Robotics and Automation,* IEEE Computer Society, Los Angeles, pp. 1772–1778.

Maciejowski, J.M. (1989). *Multivariable Feedback Design,* Addison-Wesley, Reading, MA.

Saridis, G.N. (1988). Knowledge Implementation: Structures of Intelligent Control Systems, *J. Robotic Syst.,* Vol. 5, No. 4, pp. 255–268.

Slotine, J.J.E. (1985). The Robust Control of Robot Manipulators, *Int. J. Robotics Res.,* Vol. 4, No. 2, pp. 49–64.

Young, K.K.D. (1978). Design of Variable Structure Model-Following Control Systems, *IEEE Trans. Autom. Control,* Vol. AC-23, pp. 1079–1085.

Zhou, Y. and de Silva, C.W. (1995). Adaptive Control of an Industrial Robot Retrofitted With an Open-Architechture Controller, *ASME J. Dyn. Syst. Measure. Control,* in press.

2 KNOWLEDGE REPRESENTATION AND PROCESSING

▬ INTRODUCTION

In Chapter 1 we learned that the conventional control techniques have many attractive features such as *stability* and *robustness* in the presence of external disturbances and model uncertainty, particularly when applied to *linear, time-invariant*, and *uncoupled* plants. However, these techniques can run into problems when controlling complex plants that are difficult to model. Also, the precise or "hard" (nonsoft) algorithms involved in these control methods do not have the capabilities of an *intelligent system*, particularly in terms of effectively using qualitative, fuzzy, uncertain, and incomplete information, heuristics, intuition, and of learning through experience. Furthermore, complex control algorithm can increase the "loop delay" (loop dead time) as a result of the increased processing time that would be necessary, and this may result in poor control. Another difficulty might be that when several different control techniques are available for a plant, the conventional approaches do not provide a direct way to decide when or whether to use a particular control technique, without the intervention of a human expert. We suggested that knowledge-based "soft" control could provide a suitable alternative in handling these difficulties. An intelligent controller typically consists of some form of a *knowledge base* to "represent" the knowledge that would be needed for control, and an *inference mechanism* to "process" knowledge through *reasoning*, perhaps on the basis of a set of new data, and make decisions. It follows that knowledge representation and knowledge processing are of fundamental importance in intelligent control (de Silva and MacFarlane, 1989). In this chapter we shall study several approaches to knowledge representation and processing.

■ KNOWLEDGE AND INTELLIGENCE

Every day we come across *information* in the form of, for example, newspapers, radio and TV broadcasts, sensory data, books, charts, and computer text. Information itself does not represent knowledge. We must "acquire" knowledge, e.g., through experience and by learning. In this sense information is subjected to a form of high-level processing to gain knowledge, and hence, we may interpret knowledge as *structured information*. Specialized and enhanced knowledge may be termed *expertise*. Here again, knowledge is subjected to high-level processing to acquire expertise. It is difficult to give a precise and general definition for intelligence. Instead, we often determine certain actions and behaviors to be "intelligent", depending on their outward characteristics and features. Thus, it is more appropriate to describe some important characteristics of an intelligent system (person) rather than to define intelligence itself. In particular, intelligent systems generally have a capacity to acquire and apply knowledge in a proper (intelligent!) manner and have the capabilities of *perception, reasoning, learning*, and making inferences from *incomplete information*.

According to Marvin Minsky of the Massachusetts Institute of Technology (Cambridge), *artificial intelligence* (AI) is "the science of making machines do things that would require intelligence if done by men". At the basic level of implementation, however, what a machine (computer) does is quite *procedural* and cannot be considered intelligent. In a wider perspective, and particularly viewing from outside and not from within the machine, any machine that appears to perform tasks in an intelligent manner can be considered an intelligent machine. An appropriate representation of knowledge, including intuition and heuristic knowledge, is central to the development of *machine intelligence* and of knowledge-based systems.

In a knowledge-based system two types of knowledge are needed: knowledge of the problem (problem representation or modeling) and knowledge regarding methods for solving the problem. Ways of representing and processing knowledge include (Staugaard, 1987):

1. logic
2. semantic networks
3. frames
4. production systems

We shall outline these four approaches of knowledge representation and processing.

LOGIC

In logic knowledge is represented by statements called *propositions*, which may be joined together using *connectives*. The knowledge may be processed through reasoning by the application of various laws of logic, including an appropriate *rule of inference*.

We shall limit our discussion to crisp, binary (i.e., *two-valued*) logic. In this case a proposition can assume one of only two truth values: true (T), false (F).

Example
Consider the following propositions:

1. "Charcoal is white".
2. "Snow is cold".
3. "Temperature is above 60°C".

Here, proposition 1 has a truth value F, proposition 2 has a truth value T, but for proposition 3 the truth value depends on the actual value of the temperature. If it is above 60°C the truth value is T, and otherwise it is F.

In "real life" we commonly make use of many "shades" of truth. Binary logic may be extended to many-valued logic that can handle different *grades* of truth value in between T and F. In fact, fuzzy logic may be interpreted as a further generalization of this idea, and we shall discuss this later.

We have stated that in intelligent control a knowledge base is processed through reasoning, subjected to a given set of data (measurements, observations, external commands, previous decisions, etc.) in order to arrive at new control decisions. A simple proposition does not usually make a knowledge base. Many propositions connected by logical connectives such as AND, OR, NOT, EQUALS, and IMPLIES may be needed. The *truth tables* of these connectives are shown in Table 1. Note that a truth table gives the truth values of a combined proposition in terms of the truth values of its individual propositions.

Complement

The complement of a proposition A is "NOT A" and may be denoted by symbols such as A', $\bar{A}$, and $\tilde{A}$. It is clear that when A is *TRUE*, then NOT A is *FALSE* and vice versa.

Union

The union of two propositions A and B is "A OR B" and is denoted by symbols such as $A \vee B$, $A \cup B$, and $A + B$. In this case the combined proposition is true if at least one of its constituents is true. Note that this not the "Exclusive OR," where "A OR B" is false when both A and B are true, and is true only when either A is true or B is true.

TABLE 1
TRUTH TABLES OF THE LOGICAL CONNECTIVES

(a) Complement (NOT)			(b) Union (OR)		
A	NOT *A*		*A*	*B*	*A* OR *B*
T	F		T	T	T
F	T		T	F	T
			F	T	T
			F	F	F

(c) Intersection (AND)			(d) Implication (IF-THEN)		
A	*B*	*A* AND *B*	*A*	*B*	IF *A* THEN *B*
T	T	T	T	T	T
T	F	F	T	F	F
F	T	F	F	T	T
F	F	F	F	F	T

Intersection

The intersection of A and B is "A AND B" and is denoted by symbols such as $A \wedge B$, $A \cap B$, and $A \bullet B$. In this case the combined proposition is true if and only if both constituents are true.

Implication

In intelligent control the knowledge base typically consists of a set of "IF-THEN rules" in which "implication" is the main connective used. The statement "A implies B" is the same as "IF A THEN B". This may be denoted by symbols such as $A \rightarrow B$ and $A \subset B$. The latter symbol ($\subset$) is called "set inclusion" and is particularly applicable when the two sets A and B belong to the same universe of discourse X.

Note that if both A and B are true then $A \rightarrow B$ is true. If A is false, the statement "when A is true then B is also true" is not violated regardless of whether B is true or false. However, if A is true and B is false, then the statement $A \rightarrow B$ is clearly false. These facts are represented by the truth table (d) in Table 1.

Example

We can show that in accordance with the two-valued crisp logic, the statement "IF A THEN B" is identical to "(NOT A) OR B". First, we form the truth table of (NOT A) OR B as below:

A	*A'*	*B*	$A' \vee B$
T	F	T	T
T	F	F	F
F	T	T	T
F	T	F	T

KNOWLEDGE REPRESENTATION AND PROCESSING

Here, we have used truth tables (a) and (b) in Table 1. Note that the result is identical to truth table (d) in Table 1. This equivalence is commonly exploited in logic associated with knowledge-based control. It is appropriate to mention here that the two propositions A and B are equivalent if $A \rightarrow B$ and also $B \rightarrow A$. This may be denoted by either $A \leftrightarrow B$ or $A \equiv B$.

Note that the statement "$A \leftrightarrow B$" is true either "if both A and B are true" or "if both A and B are false".

Logic Processing (Reasoning and Inference)

Thus far we have discussed the *representation* of knowledge in logic. Now let us address the *processing* of knowledge in logic. The typical objective of knowledge processing is to make inferences. This may involve the following steps:

1. Simplify the knowledge base by applying various laws of logic.
2. Substitute into the knowledge base any new information (including data and previous inferences).
3. Apply an appropriate rule of inference.

Note that these steps may be followed in any order and repeated any number of times, depending on the problem.

Laws of Logic

These laws govern the truth-value equivalence of various types of logical expressions consisting of AND, OR, NOT, etc. They are valuable in simplifying (reducing) a logical knowledge base. Consider three propositions A, B, and C, whose truth values are not known. Also, suppose that X is a proposition that is always true (T) and ϕ is a proposition that is always false (F). With this notation the important laws of logic are summarized in Table 2. It is either obvious or may be easily verified, perhaps using Table 1, that in each equation given in Table 2 the truth value of the left-hand side is equal to that of the right-hand side.

Example

Consider the two propositions:

A = "Plant 1 is stable".
B = "Plant 2 is stable".

Then consider the combined proposition:

$\overline{A \wedge B}$ = "It is not true that both plants 1 and 2 are stable".
$\overline{A} \vee \overline{B}$ = "Either plant A is not stable or plant 2 is not stable".

Obviously, both combined propositions have the same truth value. This verifies the *DeMorgan law* given in Table 2.

TABLE 2
SOME LAWS OF LOGIC

Law	Truth value equivalence
Commutativity	$A \wedge B = B \wedge A$
	$A \vee B = B \vee A$
Associativity	$(A \wedge B) \wedge C = A \wedge (B \wedge C)$
	$(A \vee B) \vee C = A \vee (B \vee C)$
Distributivity	$A \wedge (B \vee C) = (A \wedge B) \vee (A \wedge C)$
	$A \vee (B \wedge C) = (A \vee B) \wedge (A \vee C)$
Absorption	$A \vee (A \wedge B) = A$
	$A \wedge (A \vee B) = A$
Idempotency	$A \vee A = A$
	$A \wedge A = A$
Exclusion	$A \vee \overline{A} = X = T$
	$A \wedge \overline{A} = \phi = F$
DeMorgan	$\overline{A \wedge B} = \overline{A} \vee \overline{B}$
	$\overline{A \vee B} = \overline{A} \wedge \overline{B}$
Boundary conditions	$A \vee X = X = T$
	$A \wedge X = A$
	$A \vee \phi = A$
	$A \wedge \phi = \phi = F$

Rules of Inference

Rules of inference are used in reasoning associated with knowledge processing in order to make inferences. Two rules of inference are discussed here:

1. Conjunction rule of inference (CRI)
2. Modus ponens rule of inference (MPRI)

The CRI states that if a proposition A is true and a second proposition B is true, then the combined proposition "A AND B" is also true. This seemingly simple and obvious fact has application in rule-based reasoning. As an example, consider a simple knowledge base of response characteristics consisting of just two propositions:

1. "The speed of response is high".
2. "The overshoot is excessive".

Note that each statement is a proposition which may or may not be true. Now suppose that we have some new data (e.g., from a speed sensor and a position sensor) which indicate a high speed of response and also an excessive overshoot. With these data, each one of the two propositions becomes a *premise*, which is assumed to be true. Then by applying the conjunction rule of inference we can infer the following result:

"The speed of response is high and the overshoot is excessive".

The MPRI states that if proposition A is true and the implication $A \to B$ holds, it can be concluded that proposition B is also true. This rule of inference is much more widely applicable than the CRI in rule-based reasoning. To illustrate this consider the following example. Suppose that a knowledge base has the rule

"If the trajectory error exceeds 1.0 mm the robot performance is unacceptable".

Also, suppose that during operation of the robot a position sensor indicated that the trajectory error was 1.1 mm. Then, by using this information and applying the MPRI the following inference can be made:

"Robot performance is unacceptable".

Note that this inference may serve as data (new information) for another rule such as

"If the robot performance is unacceptable, then tune the controller",

which may be processed by another application of MPRI.

Propositional Calculus and Predicate Calculus

Propositional calculus is the branch of logic in which propositions (i.e., statements that are either true or false) are used in logic "calculations". Note that calculus in the present terminology implies "the approach of calculation". We have primarily discussed propositional calculus. Predicate calculus is a generalized version of propositional calculus in which, in a predicate statement, a predicate serves a purpose similar to that of a mathematical function and may be applied to a constant or variable argument. Hence, a logic statement in this case will take the form

Predicate (Argument)

The predicate describes the item within the argument. For example, the statement

Is High (Control Gain)

corresponds to *"Control gain is high"*. In fact, a more general version would be

Is High (x).

Here, a variable x is used as the argument. If x denotes "control gain", then we have precisely the previous statement. If we wish, however, x may denote something else such as "temperature".

Reasoning or (knowledge processing) in predicate calculus is quite similar to, although more general than, that in propositional

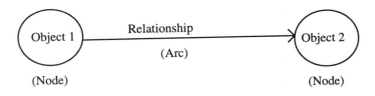

Figure 1 A segment of a semantic network.

calculus, and may be accomplished through the application of a rule of inference.

Example
Consider the knowledge base consisting of the rule

$$(\forall x)[\text{Servomotor}(x) \rightarrow \text{Speed\_Sensor}(x)]$$

Here, $\forall x$ means "for all x". The rule states that all servomotors have speed sensors and may be represented by a model number which is denoted by variable x. Now suppose that a particular positioning system has a servomotor whose model number is H0176. This is now an available data item and is given by *"Servomotor (H0176)"*. Then, by applying the MPRI to the rule base, we obtain the inference *"Speed_Sensor (H0176)"*; i.e., a speed sensor is found in that particular motor unit.

■ SEMANTIC NETWORKS

Semantic networks are used in the graphical representation and processing of knowledge. Here, knowledge objects are represented in a network, and their relationships are represented by *arcs* or lines which are directed, linking the nodes. An arc represents a particular association between two objects. For example, a marriage can be represented by a *"married to"* arc joining a *husband* object and a *wife* object. Knowledge represented by a semantic network is processed using network searching procedures, usually starting with some available data and ending with a set of conclusions. The nomenclature of a semantic network is described in Figure 1. The object at a node may be generic (e.g., "Plant") or specific (e.g., "Plant 1"). Relationships represented by links can take many forms. Examples are IS_A, HAS, CONTAINS, and HAS_FEATURE.

Example
Consider the semantic network shown in Figure 2.
This semantic net represents the following knowledge:

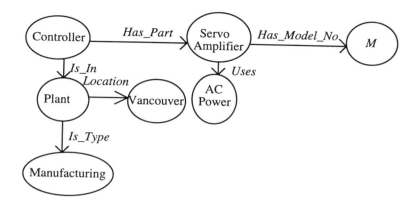

Figure 2 Semantic network.

1. The controller has a servo amp that uses AC power.
2. The controller is in a plant that is located in Vancouver.
3. It is a manufacturing plant.
4. Servo amplifiers of different models are present in the controller.

Note that M is a generic model number.

Semantic nets take a hierarchical structure when "Is_A" relationship alone is used in its representation. This is just a special case.

■ FRAMES

Frames are commonly used in AI applications for knowledge representation and processing. A frame is a data structure developed to represent *expectational knowledge*, i.e., knowledge of what to expect when entering a given situation for the first time. Common sense knowledge or general knowledge can be represented in this way, and new information (new frames) can be interpreted using old information (old frames) in a hierarchical manner. A frame may contain *context knowledge* (facts) and *action knowledge* (cause-effect relations). The frames that contain action rules (i.e., *procedural knowledge*) are called *action frames*. The other type of frames that define the object distribution (i.e., *factual knowledge*) are called *situational frames*.

A frame consists of a *frame label* and a set of *slots*, as schematically shown in Figure 3. Each slot stores either a set of statements (rules) or another frame that is at the level below the present level in the hierarchy. In this manner knowledge may be represented at different hierarchical levels in different degrees of detail (i.e., *resolution*).

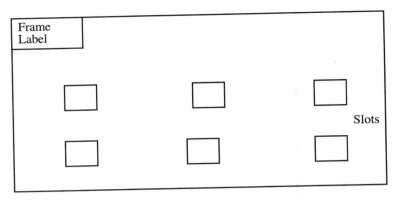

Figure 3 Semantic representation of a frame.

Generation of a knowledge base using frames is an evolutionary procedure. First, a generic frame, (i.e., a *template* or a *schematic frame*), is activated within the knowledge-base computer. This schematic frame will have a default label and a standard number of default sets. Subsequent steps are as follows:

1. Give a name (label) to the frame.
2. Name the slots to represent objects contained within the frame.
3. Open the slots one at a time, and either program the set of action rules or define the lower level frames, as in step 2.
4. Repeat step 3 as many times as required, depending on the required number of levels in the hierarchy.

Example

Consider a frame representing a manufacturing workcell, as shown in Figure 4. Note that the workcell consists of a four-axis robot, a milling machine, a conveyor, a positioning table, an inspection station, and a cell-host computer. They are represented by slots in the cell frame. Each of these slots contains a situational frame. For example, the robot slot stores the robot frame as shown. The cell host also has a data slot which stores the data files that are needed for the operation of the cell. The programs slot contains an action frame which carries various procedures needed in the operation of the cell. For example, the programs needed for carrying out various cell tasks through proper coordination of the cell components, monitoring, and control, are carried by the "Host Programs" frame.

So far we have considered knowledge representation using frames. Now let us address reasoning or knowledge processing, which is essential for problem solving through frames. Reasoning using frames is done as follows: an external input (e.g., operator command or a sensor signal) or a program action triggers a frame according to some heuristics; then, slots in the frame are matched with *context* (i.e.,

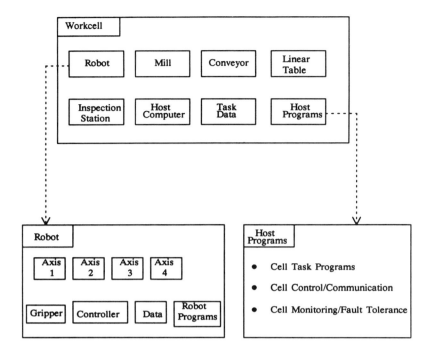

Figure 4 Representation of a workcell using frames.

current data, including sensory inputs), which may lead to a possible updating of slot values in the current frame and to triggering of subsequent frames once the conditions are matched. Finally, the appropriate procedures are executed by the action frames.

As an example, again consider the manufacturing workcell problem shown in Figure 4. During operation of the reasoning mechanism, the knowledge-base computer opens a workcell frame. There will be manual or automatic data input to the computer indicating the cell components (e.g., robot, mill, etc.). The system matches the actual components with the slots of the frame. If no match can be found and if the system has another workcell frame, that frame is matched. In this manner the closest match is chosen and updated to obtain an exact match for all frames at all levels. Then the action slots of the frame at the highest level (cell level) are activated. This will initiate the cell actions, which in turn trigger the cell component actions. Note that in a hierarchical system of this type, the knowledge and actions needed at a higher level are generally more "intelligent" than those needed at a lower level; however, the resolution (i.e., degree of detail) of the information needed at a higher level is lower than at a lower level (de Silva, 1991). For example, the coordination of a multicomponent workcell is a far more intelligent activity than the

34 INTELLIGENT CONTROL: FUZZY LOGIC APPLICATIONS

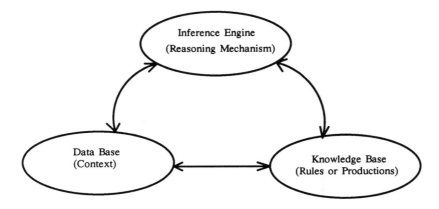

Figure 5 The structure of a production system.

servo control of an axis of a robot, but the information for the former is needed at a considerably lower bandwidth and resolution than that for the latter.

■ PRODUCTION SYSTEMS

Production systems or *rule-based systems* are commonly used in knowledge representation and processing (problem solution) associated with artificial intelligence. An expert system is one example of a production system. The structure of a typical production system is shown in Figure 5. Here, knowledge is represented by a set of *rules* (or *productions*) stored in the *knowledge base*. The *data base* contains the current data (or context) of the process. The *inference engine* is the reasoning mechanism which controls rule matching, coordinates and organizes the sequence of steps used in solving a particular problem, and resolves any conflicts.

Knowledge representation using a set of IF-THEN rules is not an unfamiliar concept. For example, a maintenance or troubleshooting manual of a machine (e.g., automobile) has such rules, perhaps in tabular form. Also, a production system may be used as a simple model for human reasoning; sensory data fire rules in the short-term memory which will lead to the firing of more complex rules in the long-term memory.

The operation (processing) of a production system proceeds as follows. Some new data are generated (e.g., from sensors or external commands) and stored in appropriate locations in the data base of the system. This is the new context. The inference engine tries to match the new data with the *condition part* (i.e., the IF part or the *antecedent*) of the rules in the knowledge base. This is called *rule searching*. If the condition part of a rule matches the data, that rule is "fired", which generates an action dictated by the *action part* (i.e., the THEN part or the *consequent*) of the rule. In fact, firing of a rule amounts to the generation (inference) of new facts, and this in turn may form a context that will lead to the satisfaction (firing) of other rules.

Example
Consider a knowledge base for selecting a control technique, as given by the following set of rules:

1. If the plant is linear and uncoupled, then use Control Category A.
2. If the plant is linear and coupled, then use Control Category B.
3. If the plant is nonlinear, then use Control Category C.
4. If Category A, and a plant model is known, then use Subgroup 1.
5. If Category B, and a plant model is known, then use Subgroup 2.
6. If Subgroup 2, and high model uncertainty, then use H_∞ control.

Now suppose that the data base received the following context:

Linear
Coupled
Model available
Model uncertainty high

In this case the first two items in the context will fire rule 2. The results will form a new item in the context. This new item, along with the third item in the old context, will fire rule 5. The result, along with the last item of the old context, will fire rule 6, which will lead to the selection of H_∞ control.

Reasoning Strategies
Two strategies are available to perform reasoning and make inferences in a production system:

1. Forward Chaining
2. Backward Chaining

Forward chaining is a *data-driven* search method. Here, the rule base is searched to match an IF (condition) part of a rule with known facts or data (context), and if a match is detected that rule is fired (i.e., the THEN part or ACTION part of the rule is activated). Obviously, this is the direct strategy and is a *bottom-up* approach. Actions could include creation, deletion, and updating of data in the data base. As we saw in the previous example one action can lead to firing of one or more new rules. The inference engine is responsible for sequencing the matching (searching) and action cycles. A production system that uses forward chaining is termed a *forward production system* (FPS). This type of system is particularly useful in knowledge-based control. In the previous example on control-technique selection we tacitly assumed forward chaining.

In backward chaining, which is a *top-down* search process, a hypothesized conclusion is matched with rules in the knowledge base in order to determine the context (facts) that supports the particular conclusion. If enough facts support a hypothesis, the hypothesis is accepted. Backward chaining is useful in situations which call for diagnosis, theorem proving, and generally in applications in which a logical explanation must be attached to each action. A production system which uses backward chaining is called a *backward chaining system* (BCS). In the previous example on control-technique selection, if we had hypothesized the selection of another control technique (e.g., the linear quadratic Gaussian method) and proceeded backward, searching the rule base for the context information that satisfies this hypothesis, we would arrive at a set of requirements (premises). However, they would not match the entire context of the situation (specifically, *model uncertainty high* will not be matched), and hence the hypothesis would be false. Then we should try another hypothesis and proceed similarly until a suitable match is obtained.

Conflict Resolution Methods

When the context data are matched with the condition parts of the rules in a rule base, it may be possible that more than one rule is satisfied. The set of rules satisfied in this manner is termed the *conflict set*. Then, a method of conflict resolution must be invoked to select the rule that would be fired. Methods of conflict resolution include the following:

1. First match
2. Toughest match
3. Privileged match
4. Most recent match

In the first method the very first rule that is satisfied during searching will be fired. This is a very simple strategy, but may not produce the best performance in general. In the second method the rule with the most condition elements within the conflict set will be fired. For example, if the conflict set has the following two rules:

> If the temperature is high, then increase the coolant flow rate.
> If the temperature is high and the coolant flow rate is maximum, then shut down the plant.

Here, the toughest match is the second rule.

The rules in a rule base may be assigned various weightings and priorities depending on their significance. The privileged match is the rule with the highest priority in the conflict set. For example, a priority may be assigned in proportion to the toughness of the match.

In this case methods 2 and 3 are identical. Alternatively, a priority may be assigned to a rule on the basis of the significance or consequences of its actions.

The most recent match is the rule in the conflict set whose condition part satisfied the most recent entries of data. In this method a higher priority is given to more recently arrived data in the data base.

Expert Systems

Expert systems are a class of production systems which are typically used in a *consultation mode* and provide an alternative to consulting a human expert. Expert systems have been developed for a variety of applications such as medical diagnosis and prescription of treatment, mineral exploration, financial advising, legal consultation, and system troubleshooting and maintenance. The development of an expert system requires one or preferably more human experts in the field of interest and a *knowledge engineer*. The experts will be involved in defining the problem domain to be solved and in providing the necessary expertise and knowledge that will lead to the solution of a class of problems in that domain. The knowledge engineers are involved in acquiring and representing the available knowledge and in the overall process of programming and debugging the system. Experts and knowledge engineers both will be involved in testing and evaluating the system.

In developing the knowledge base of an expert system the experts involved may rely on published information, experimental findings, and any other available information in addition to their own knowledge. Testing and evaluation may involve trials in which the experts pose problems to the expert system and evaluate the inferences made by the system. Many of these test problems may have already-known solutions. Some others may not be so clear cut, and hence, the responses of the expert system should be carefully scrutinized by the experts during the testing and evaluation phases. The tests should involve the ultimate users of the system as well. In this regard, the features of the user interface including simplicity, use of a common language, graphics, and voice facilities are important.

Note that an expert system may be equally useful to both an expert and a layperson. For example, it is difficult for one expert to possess a complete knowledge in all aspects of a problem, and the solutions can be quite complex. The expert may turn to a good expert system which will provide solutions that the expert could evaluate further, perhaps with the assistance of other experts, before adopting. Ideally, an expert system should have an "evolving" knowledge base which has the capability to learn and continuously update its knowledge base.

■ SUMMARY

Knowledge is *structured information*, and knowledge acquisition is done through *learning* and *experience*, which are forms of *high-level processing*. Expertise is enhanced or *specialized knowledge*. Intelligence is characterized by the distinctive abilities of an intelligent system, which include the capabilities of *perception, reasoning, learning*, making inferences from *incomplete information*, and the capacity to acquire and apply knowledge properly.

Knowledge representation and processing are the key to any knowledge-based (intelligent) system. In *two-valued logic* knowledge is represented by *propositions*, which may be joined by the logical connectives such as AND, OR, NOT, and IMPLIES. A proposition may have a truth value of *true* (T) or *false* (F). Knowledge is processed through the *reasoning* method in order to make *inferences*. In logic, knowledge is processed through various laws, including *rules of inference*. Predicate calculus is a general form of propositional calculus and represents the mathematics of making "calculations" in logic.

Semantic networks are graphic structures in which objects, represented by *nodes*, are joined by *arcs*, which represent relationships between the objects. The knowledge represented by a semantic network is processed by first assigning some known data or facts to the corresponding nodes and then reasoning through a network searching procedure. A semantic network that exclusively employs the "Is-A" relationship has a characteristic hierarchical structure.

Use of *frames* for knowledge representation and processing is common in AI applications. This provides a natural, hierarchical structure for knowledge representation and reasoning. A frame has a *label* and a set of *slots*. Each slot contains either another frame (hence the hierarchy) or a set of *procedures*. For knowledge representation, one starts with a *template frame* (*schema*) and systematically customizes it by matching, adding, and deleting slots and then appropriately filling these slots with either *situational frames* (with *factual knowledge*) or *action frames* (with *procedural knowledge*). Knowledge processing is done by triggering a frame through an input (e.g., program action, sensor signal, command), filling the context data, and executing the procedures dictated by the action frame.

A production system is a *rule-based system* for representing and processing knowledge. It consists of a *knowledge base* in the form of a set of IF-THEN rules, a *data base* which stores the context data, and an *inference engine* which performs reasoning. Here, the inference engine searches the rule base and matches the *context* with the "condition" part of a rule. The matched rule is "fired", thereby performing the *action* dictated by the rule. The action may include

KNOWLEDGE REPRESENTATION AND PROCESSING

creating, deleting, and updating data (context) in the data base. A given context may satisfy more than one rule. The set of satisfied rules is called the *conflict set*, and choosing one rule from the set, for firing, is called *conflict resolution*. Conflict resolution may be achieved by several methods, including the selection of first match, toughest match, privileged match, and most recent match. Either *forward chaining* or *backward chaining* may be used for reasoning and inference making in a production system. Forward chaining is more direct where reasoning proceeds in the forward direction by sequentially satisfying conditions and inferring the corresponding actions, starting with a set of known data. In backward chaining first a particular conclusion is hypothesized and then one proceeds backwards. If the inferred context matches the actual context, the hypothesis is true. Otherwise an alternative hypothesis is used and proceeded until a match is obtained.

An expert system is a production system that serves the purpose of a human expert in a given *domain of expertise*. An expert system is developed by a group consisting of *domain experts* and *knowledge engineers*. The experts define the problems and provide knowledge through their own expertise and available knowledge. The knowledge engineers acquire and represent this knowledge and program it into the system. The experts and the knowledge engineers cooperate in testing and evaluating the system.

An expert system is used in the consultation (dialogue) mode, and hence a good *user interface* is essential. The users of an expert system may include experts themselves who need to get a "second opinion" on a complex problem or laypersons who need expert advice on a difficult problem. Applications include systems for providing medical, legal, business, and technical advice.

■ PROBLEMS

1. In statistical process control (SPC), decisions as to whether a process is in control are made using a control chart. A control chart simply consists of two lines, the *lower control line* and the *upper control line*, drawn parallel to the time axis, and are in units of the controlled (response) variable, as shown in Figure 6. During operation of the plant, the variable (response) of interest is sampled at intervals, and a sample average is computed. If this value remains within the two control lines, the process is in control. In this case the process error that exists is caused by inherent random causes and can be neglected. If the control limits are exceeded, appropriate control actions are taken to correct the

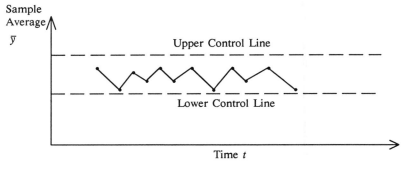

Figure 6 A control chart.

situation. Do you consider SPC an intelligent control technique? Explain.

2. If "$A \rightarrow B$" is F and "$B \rightarrow A$" is T, what is the truth value of "$(A \rightarrow B)$ AND $(B \rightarrow A)$"? Using this result and the truth table of "$A \rightarrow B$" determine the truth table of "$A \leftrightarrow B$".

3. Consider the following knowledge base:
   The step response of the robot has large oscillations.
   If a plant response is oscillatory, then increase the derivative action.
   What rule of inference may be used in processing this and what is the resulting inference? Indicate the steps involved in the reasoning process.

4. Consider the knowledge base:
   LOW means <16°C.
   SLIGHT means one division.
   If the temperature is LOW, then SLIGHTLY increase the thermostat setting.
   Is this a fuzzy knowledge base? Explain

5. Which method of knowledge representation and processing given below would be the most appropriate model for human reasoning?
   (a) Two-valued logic
   (b) Semantic networks
   (c) Frames
   (d) Production systems

6. Define the term "information resolution".

   In a hierarchical control system the degree of intelligence needed for various actions generally increases with the hierarchical level.

Also, the information resolution generally decreases with the hierarchical level. Considering the example of a manufacturing workcell, verify these relationships.

7. In conventional computer programs, data, instructions, and reasoning are all integrated, but the knowledge-based application programs are usually "object-oriented", in which these three items are separately structured and located. Give an advantage of this separation.

8. One definition of an expert system is "a computer program that embodies expert knowledge and understanding about a particular field of expertise, which can be used to assist in the solution of a problem in that field". List several areas in which expert systems are found to be useful. Choose a specific application and develop a simple rule base that may be used in a knowledge-based system.

9. Expert systems should be able to not only provide answers to questions but also provide explanations for, or reasoning behind, these answers. Furthermore, they should have the capability to consider alternative goals, not just the single goal that is implied by a specific question. For example, suppose that an expert system for condition monitoring of a workcell is asked the question, "Is the performance acceptable?" With *backward chaining*, a response of "Yes" might be generated. This response might be true but would not be adequate. Explore this, and indicate what other details could be expected from the expert system.

10. A difficult and important task in the development of any knowledge-based (or, artificial-intelligence) application is the process of *knowledge engineering*. This concerns the acquisition of expert knowledge and representation of the knowledge in a form that is compatible with the specific knowledge-based system. What are some other important considerations in developing such an application?

11. Critically evaluate the statement "Power of an expert system is derived from its knowledge base, not from the particular formalism and inference scheme it employs".

12. The decision tree shown in Figure 7 provides a simple knowledge base for climate control of a room. Translate this into a rule base.

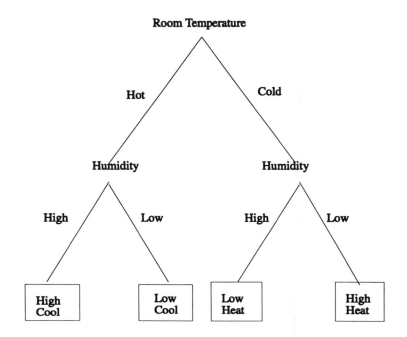

Figure 7 The decision tree of a climate control system.

REFERENCES

de Silva, C.W. (1991). Fuzzy Information and Degree of Resolution within the Context of a Control Hierarchy, *Proc. IEEECON'91*, IEEE, New York, pp. 1590–1595.

de Silva, C.W. and MacFarlane, A.G.J. (1989). *Knowledge-Based Control with Application to Robots*, Springer-Verlag, Berlin.

Staugaard, A.C. (1987). *Robotics and AI*, Prentice-Hall, Englewood Cliffs, NJ.

3 FUNDAMENTALS OF FUZZY LOGIC

INTRODUCTION

Knowledge-based control relies on some method of representing and processing knowledge. In Chapter 2, we introduced several methods that may be employed for this purpose. In particular, *bivalent logic* was introduced and some of its properties were introduced. This two-state, crisp logic uses only two quantities, *true* (T) and *false* (F) as truth values.

Real-life situations do not always assume two crisp states T and F, with a crisp line dividing them. Also, linguistic descriptors such as "fast", "warm", and "large" are not crisp quantities and tend to be quite subjective and qualitative. Thus, a statement such as "the water is warm" may have a truth value that is partially true and partially false. Conventional, bivalent logic is inadequate to handle such situations. Conventional logic introduces some well-known paradoxes. For example, the statement, "all my statements are lies" does not have a crisp truth value, and is hence not amenable to conventional, two-state logic. This statement, if assumed to be true, will be false, and vice versa. Fuzzy logic is a quite general form of logic that can deal with many such noncrisp situations. However, it may violate some of the laws of bivalent logic, as given in Table 2 of Chapter 2. In particular, the *law of exclusion* will not be exactly satisfied in fuzzy logic (Kosko, 1992).

This chapter presents the basic theory of fuzzy logic, which uses the concept of fuzzy sets. Compositional rule is what is applied for making inferences in decision making associated with fuzzy logic. Principles underlying the *compositional rule of inference* (Zadeh, 1973; 1979) are systematically developed in this chapter.

43

■ FUZZY SETS

Fuzzy logic is the logic that deals with fuzzy sets. The concepts of fuzzy sets and fuzzy logic are used in fuzzy control. The rule of fuzzy inference, namely, the *compositional rule of inference*, is particularly applicable in the context of fuzzy control. In this section the mathematics of fuzzy sets is summarized, as this is necessary to understand the theory of fuzzy control. What is presented is an interpretation of some theoretical considerations described primarily by Dubois and Prade (1980). Geometrical illustrations and examples have been added in order to facilitate the interpretation.

A fuzzy set is a set that does not have a sharp (crisp) boundary. In other words, there is a softness associated with the membership of elements in a fuzzy set. Consider a *universe of discourse X* whose elements are denoted by x. A fuzzy set A in X may be represented by a Venn diagram, as in Figure 1. Generally, the elements x are not numerical quantities, but for analytical convenience the elements x are assigned real numerical values.

Note that a variable (e.g., temperature) can take a fuzzy value (e.g., "high"). A fuzzy value (or *fuzzy label*) may be represented by a fuzzy set. The membership in a fuzzy set may be represented by a membership function, as explained below.

Membership Function

A fuzzy set may be represented by a membership function. This function gives the grade (degree) of membership within the set, of any element of the universe of discourse. The membership function maps the elements of the universe onto numerical values in the interval [0, 1]. Specifically,

$$\mu_A(x): X \rightarrow [0, 1] \tag{1}$$

where $\mu_A(x)$ is the membership function of the fuzzy set A. Note that a membership function is a possibility function and not a probability function. A membership function value of zero implies that the corresponding element is definitely not an element of the fuzzy set. A membership function value of unity means that the corresponding element is definitely an element of the fuzzy set. A grade of membership greater than 0 and less than 1 corresponds to a noncrisp (or fuzzy) membership, and the corresponding elements fall on the fuzzy boundary of the set. A typical membership function is shown in Figure 2.

Symbolic Representation

A universe of discourse and a membership function that spans the universe completely defines a fuzzy set. We are not usually interested

FUNDAMENTALS OF FUZZY LOGIC 45

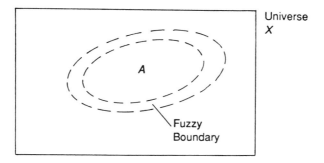

Figure 1 The Venn diagram of a fuzzy set.

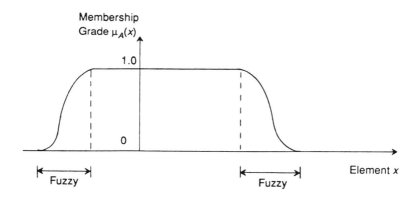

Figure 2 The membership function of a fuzzy set.

in the elements with zero grade of membership. The crisp set formed by leaving out all the boundary elements (i.e., leading and trailing elements in the membership function) that have a zero membership grade and by retaining all the remaining elements is termed the *support set* of the particular fuzzy set. The support set is a crisp subset of the universe. Furthermore, a fuzzy set is clearly a subset of its support set. A fuzzy set may be symbolically represented as

$$A = \{x \mid \mu_A(x)\} \tag{2}$$

If the universe is discrete with elements x_i, then a fuzzy set may be specified using a convenient *form of notation* due to Zadeh, in which each element is paired with its grade of membership in the form of a "formal series" as

$$A = \mu_A(x_1)/x_1 + \mu_A(x_2)/x_2 + \cdots + \mu_A(x_i)/x_i + \cdots$$

or

$$A = \sum_{x_i \in X} \frac{\mu_A(x_i)}{x_i} \qquad (3)$$

If the universe is continuous, an equivalent form of notation is given in terms of a *symbolic* integration

$$A = \int_{x \in X} \frac{\mu_A(x)}{x} \qquad (4)$$

It is important to emphasize that both the series in Equation (3) and the integral in Equation (4) are symbolic shorthand forms of notation; to highlight this, no differential symbol $d(.)$ is used in Equation (4).

Three examples are given to illustrate various types of fuzzy sets and their representations.

Example 1
Suppose that the universe of discourse X is the set of positive integers (or natural numbers). Consider the fuzzy set A in this discrete universe, given by the Zadeh notation:

$$A = 0.2/3 + 0.3/4 + 1.0/5 + 0.2/6 + 0.1/7$$

This set may be interpreted as a fuzzy representation of the integer 5.

Example 2
Consider the continuous universe of discourse X representing the set of real numbers ($\Re$). The membership function:

$$\mu_A(x) = 1/[1 + (x - a)^{10}]$$

defines a fuzzy set A whose elements x softly represent those satisfying the crisp relation $x = a$. This fuzzy set corresponds to a fuzzy relation.

Example 3
Linguistic terms such as "tall men", "beautiful women", "fast cars", and "slight increase" are fuzzy labels or fuzzy descriptors which can be represented by fuzzy sets. Note that their membership in a set is subjective and somewhat vague or noncrisp.

FUZZY LOGIC OPERATIONS

It is well known that the "complement", "union", and "intersection" of crisp sets correspond to the logical operations NOT, OR, and AND, respectively, in the corresponding crisp, bivalent logic. These concepts were discussed in Chapter 2. Furthermore, we noticed that the union of a set with the complement of a second set represents an "implication" of the first set by the second set. Set inclusion (subset) is a special case of implication in which the two sets belong to the same universe. These logical operations (connectives) must be extended to fuzzy sets for use in fuzzy reasoning and fuzzy logic control. In fuzzy logic these connectives must be expressed in terms of the membership functions of the sets which are operated on.

Complement (NOT)
Consider a fuzzy set A in a universe X. Its complement A' is a fuzzy set whose membership function is given by

$$\mu_{A'}(x) = 1 - \mu_A(x) \quad \text{for all } x \in X \tag{5}$$

The complement corresponds to a NOT operation in fuzzy logic.

Union (OR)
Consider two fuzzy sets A and B in the same universe X. Their union is a fuzzy set $A \vee B$. Its membership function is given by

$$\mu_{A \vee B}(x) = max[\mu_A(x), \mu_B(x)] \quad \forall x \in X \tag{6}$$

The union corresponds to a logical OR operation. The rationale for the use of *max* is that, because the element x may belong to one set or the other, the larger of the two membership grades should apply.

Intersection (AND)
Again, consider two fuzzy sets A and B in the same universe X. Their intersection is a fuzzy set $A \wedge B$. Its membership function is given by

$$\mu_{A \wedge B}(x) = min[\mu_A(x), \mu_B(x)] \quad \forall x \in X \tag{7}$$

The intersection corresponds to a logical AND operation. The rationale for the use of *min* is that because the element x must belong to both sets simultaneously, the smaller of the two membership grades should apply.

Implication (IF–THEN)

Consider a fuzzy set A in a universe X and a second fuzzy another universe Y. The fuzzy implication $A \rightarrow B$ is a fuzzy in the *Cartesian product space* $X \times Y$. There are several int tions for this relation. Two commonly used relations for obtaining the membership function of the fuzzy implication are given below.

Method 1:

$$\mu_{A \rightarrow B}(x,y) = min[(\mu_A(x), \mu_B(y)]$$

$$\forall x \varepsilon X, \quad \forall y \varepsilon Y \qquad (8)$$

Method 2:

$$\mu_{A \rightarrow B}(x,y) = min[1, \{1 - \mu_A(x) + \mu_B(y)\}]$$

$$\forall x \varepsilon X, \quad \forall y \varepsilon Y \qquad (9)$$

Note that the first method gives an expression which is symmetric with respect to A and B. This is not intuitively satisfying because "implication" is not a commutative operation. In practice, however, this method provides good, robust results. The second method has an intuitive appeal because in crisp logic, $A \rightarrow B$ has the same truth table as [(NOT A) OR B] and hence are equivalent, as discussed in Chapter 2. Note that in Equation 9, the membership function is upper bounded to 1 using the *bounded sum* operation, as required by definition. (A membership grade cannot be greater than 1.) The first method is more commonly used because it is simpler to use and often provides more accurate results than the second method.

■ SOME DEFINITIONS

This section introduces the concepts of set inclusion. Next, we present a numerical measure for fuzziness of a set. Subsequently, the definition of α-cut of a fuzzy set is given. The section is concluded by discussing the *representation theorem*.

Set Inclusion ($A \subset B$)

The concept of subset (or, a set "included" in another set) in crisp sets may be conveniently extended to the case of fuzzy sets. Specifically, a fuzzy set A is considered a subset of another fuzzy set B, in universe X, if and only if

$$\mu_A(x) \leq \mu_B(x) \quad \text{for all } x \in X \tag{10}$$

This is denoted by $A \subset B$. Set inclusion is a special case of the set equality $A = B$ because the latter is satisfied if and only if $\mu_A(x) = \mu_B(x)$ for all $x \in X$. If $A \subset B$ and $A \neq B$, then A is called a *proper subset* of B.

Measure of Fuzziness

If the membership grade of an element is close to unity, the element is almost definitely a member of the set. Conversely, if the membership grade of an element is close to zero, the element is nearly outside the set. From this observation it should be clear that the membership of an element x in a set A is most fuzzy when the membership grade $\mu_A(x) = 0.5$. Accordingly, the fuzziness of a set may be measured by the closeness of its elements to the membership grade of 0.5. In view of this an appropriate measure of fuzziness of set A would be the inverse of

$$c_A = 2 \int_{x \in X} |\mu_A(x) - 0.5| \, dx$$

$$= \int_{x \in X} |2\mu_A(x) - 1| \, dx$$

$$= \int_{x \in X} |\mu_A(x) - \mu_{\bar{A}}(x)| \, dx \tag{11}$$

Alternatively, the fuzziness may be represented by

$$f_A = \int |\mu_A(x) - \mu_{A_{1/2}}(x)| \, dx \quad \text{where } \mu_{A_{1/2}} \text{ is the 1/2 cut of } \mu_A(x)$$

which is defined next.

α-Cut of a Fuzzy Set

The α-cut of a fuzzy set A is the crisp set formed by elements A whose membership grade is greater than or equal to a given value α. This is denoted by A_α. Specifically,

$$A_\alpha = \{x \in X | \mu_A(x) \geq \alpha\} \tag{12}$$

in which X is the universe to which A belongs (Dubois and Prade, 1980; Pedrycz, 1989).

Representation Theorem

It should be intuitively clear that a fuzzy set A may be recomposed or represented by means of all α-cuts of the fuzzy set. The representation theorem expresses this and is given by

$$\mu_A(x) = \sup_{\alpha \in [0,1]}[\alpha \mu_{A_\alpha}(x)] \tag{13}$$

Proof

Suppose that at $x = x_1$, $\mu_A(x) = \alpha_1$. Then, for $\alpha < \alpha_1$ we have $\mu_{A_\alpha}(x_1) = 1$ and for $\alpha > \alpha_1$ we have $\mu_{A_\alpha}(x_1) = 0$. It follows that for $x = x_1$, the R.H.S. of Equation (13) is equal to α_1. Q.E.D.

■ FUZZY RELATIONS

Consider two universes and X_1, = $\{x_1\}$ and X_2, = $\{x_2\}$. A crisp set R consisting of a subset of ordered pairs (x_1, x_2) is a crisp relation in the Cartesian product space $X_1 \times X_2$. An example is shown in Figure 3(a). Analogously, a fuzzy set R consisting of a subset of ordered pairs (x_1, x_2) is a fuzzy relation in the Cartesian product space $X_1 \times X_2$. The relation R may be represented by the membership function $\mu_R(x_1, x_2)$. An example of a fuzzy relation is shown in Figure 3(b). This concept can be extended in a straightforward manner to fuzzy relations in the n-dimensional Cartesian space $X_1 \times X_2 \times \cdots X_n$.

Example

Consider the fuzzy set R in the universe $X_1 \times X_2$ given by the membership function:

$$\mu_R(x_1, x_2) = 1/[1 + 100(x_1 - 3x_2)^4]$$

This is a fuzzy relation vaguely representing the crisp relation $x_1 = 3x_2$. Particularly note that all elements satisfying $x_1 = 3x_2$ have unity grade of membership and hence they are definitely in the set R. Elements satisfying, for example, $x_1 = 3.1x_2$ have membership grades less than 1; their membership in R is vague or soft. The further away the elements are from the straight line $x_1 = 3.1x_2$, the softer the membership of those elements in R.

Cartesian Product of Fuzzy Sets

Consider a crisp set A_1 in the universe X_1 and a second crisp set A_2 in another universe X_2. The Cartesian product $A_1 \times A_2$ is a subset of the Cartesian product space $X_1 \times X_2$ defined in the usual manner as shown in Figure 4(a). Next consider a fuzzy set A_1 in the universe X_1

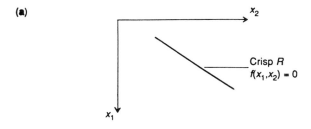

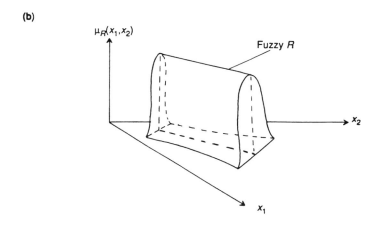

Figure 3 (a) A crisp relation R in a two-dimensional space (plane). (b) A fuzzy relation R in a two-dimensional space.

and a second fuzzy set A_2 in another universe X_2. The Cartesian product $A_1 \times A_2$ is then a fuzzy subset of the Cartesian space $X_1 \times X_2$. Its membership function is given by

$$\mu_{A_1 \times A_2}(x_1, x_2) = min[\mu_{A1}(x_1), \mu_{A2}(x_2)]$$

$$\forall x_1 \varepsilon X_1, \forall x_2 \varepsilon X_2 \qquad (14)$$

Note that the *min* combination applies here because each element (x_1, x_2), in the Cartesian product is formed by taking both elements x_1, x_2 and together, not just one or the other. An example of a Cartesian product of two fuzzy sets is shown in Figure 4(b).

The Cartesian product of two fuzzy sets is a fuzzy relation, and is identical to the first method of fuzzy implication given by Equation (8). The concept of Cartesian product can be directly extended to more than two fuzzy sets.

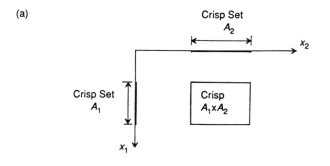

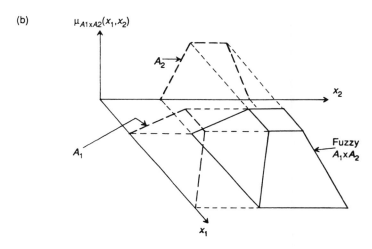

Figure 4 Cartesian product ($A_1 \times A_2$) relation of (a) two crisp sets; (b) two fuzzy sets.

Extension Principle

The extension principle was introduced by Zadeh to give a method for extending standard (non-fuzzy or crisp) mathematical concepts to their fuzzy counterparts. Consider the relation:

$$y = f(x_1, x_2, \cdots, x_r) \qquad (15)$$

where x_i are defined in the universes X_i, $i = 1, 2, \ldots, r$, and y is defined in the universe Y. This relation generally is a many-to-one mapping from the r-dimensional Cartesian product space $X_1 \times X_2 \times \cdots \times X_r$ to the one-dimensional space Y:

$$f : X_1 \times X_2 \times \cdots \times X_r \to Y \qquad (16)$$

FUNDAMENTALS OF FUZZY LOGIC

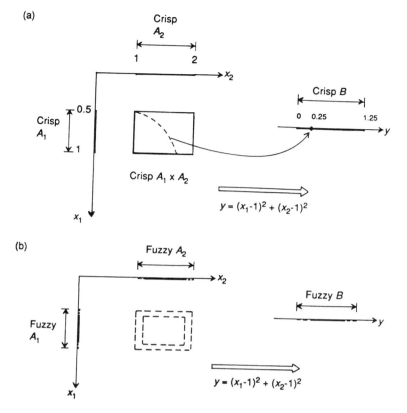

Figure 5 A mapping from a product space to a line. (a) An example of crisp sets. (b) An example of fuzzy sets (extension principle).

Suppose that the elements x_i are restricted to crisp subsets A_i of X_i. Then the relation f maps elements $(x_1, x_2, \cdots, x_r)$ within the Cartesian product $A_1 \times A_2 \times \cdots \times A_r$ onto a crisp set B, which is a subset of Y. An example is given in Figure 5(a). Note that this is a many-to-one mapping. For instance, the entire quarter circle with center (1, 1) and radius 0.5 in the product space $A_1 \times A_2$ is mapped on to the single point 0.25 on the Y line.

This idea can be extended to the case in which the sets are fuzzy. The extension principle provides the means of doing this. Note that f is still a crisp relation. According to the extension principle, the fuzzy set B to which the elements y belong has a membership function given by

$$\mu_B(y) = sup\{min[\mu_{A1}(x_1), \mu_{A2}(x_2), \cdots, \mu_{Ar}(x_r)]\}(x_1, x_2, \cdots, x_r)$$
$$y = f(x_1, x_2, \cdots, x_r) \qquad (17)$$

Note that the *min* operation is applied first because the relation among A_i is the Cartesian product (see Equation (14)). The *supremum* is applied over the mapping onto B because more than one combination of $(x_1, x_2, \cdots, x_r)$ in the fuzzy space $A_1 \times A_2 \times \cdots \times A_r$ will be mapped to the same element y in the fuzzy set B, and the most possible mapping is the one with the highest membership grade. An example of a mapping from a two-dimensional fuzzy space $A_1 \times A_2$ onto a one-dimensional fuzzy set B is shown in Figure 5(b). In the one-dimensional case of $y = f(x)$ the extension principle is given by

$$\mu_{f(A)}(y) = \sup_{x \in X}[\mu_A(x)] \tag{18}$$

Fuzzy Dynamic Systems

Consider a nonlinear and time-invariant dynamic system expressed in the discrete state-space form:

$$x_{n+1} = f(x_n, u_n) \tag{19}$$

$$y_n = g(x_n) \tag{20}$$

For the purpose of explaining the underlying concepts, only the scalar case is considered. The state variable x and the output variable y are assumed to be fuzzy, and the input variable u is assumed to be crisp. Generally, in a fuzzy system the state transition relation f and the output relation g both will be fuzzy relations with fuzzy sets F and G. Suppose that the fuzzy sets corresponding to x_n, x_{n+1}, and y_n are $X(n)$, $X(n+1)$, and $Y(n)$, respectively. A typical simulation objective for a fuzzy dynamic system would be to determine the membership functions of $X(n+1)$ and $Y(n)$ once the membership functions of $X(n)$, F, and G are known. The extension principle is applicable here, allowing for the fact that the relations f and g are also fuzzy. Specifically, we have the following results:

$$\mu_{X(n+1)}(x_{n+1}) = \sup_{x_n \in X} min\{(\mu_{X(n)}(x_n), \mu_F(x_{n+1}, x_n, u_n))\} \tag{21}$$

$$\mu_{Y(n)}(y_n) = \sup_{x_n \in X}\{min(\mu_{X(n)}(x_n), \mu_G(x_n, u_n))\} \tag{22}$$

Note that X denotes the universe in which the state variable x lies; in this case the state space. It is clear from the next section that "composition" operation is used in the above two results. Specifically, the fuzzy set $X(n+1)$ is obtained through composition of the fuzzy set $X(n)$ and the fuzzy relation F. Similarly, the fuzzy set $Y(n)$ is obtained through *composition* of the fuzzy set $X(n)$ and the fuzzy relation G. In Equation (21), for example, *min* is applied first because x_n element and relation f are matched through an AND operation. Next, *sup* is applied because many elements x_n might be mapped to the same element x_{n+1} and hence, the most desirable (possible) mapping must be chosen. This is exactly the idea of composition.

■ COMPOSITION AND INFERENCE

Approximate reasoning is used in fuzzy inference and control. In particular, the *compositional rule of inference* is utilized. We have already introduced the concept of *fuzzy implification*. We now introduce the terms *projection*, *cylindrical extension*, and *join*, which lead to the concept of *composition*. Finally, the compositional rule of inference is discussed, incorporating all these ideas.

Projection

Consider a fuzzy relation R in the Cartesian product space $X_1 \times X_2 \times \cdots \times X_n$. Suppose that the n indices are arranged as follows:

$$\{1, 2, \cdots, n\} \to \{i_1, i_2, \cdots, i_r, j_1, j_2, \cdots, j_m\} \qquad (23)$$

Note that $r + m = n$ and that i and j denote the newly ordered set of n indices. The projection of R on the subspace $X_{i1} \times X_{i2} \times \cdots \times X_{ir}$ is denoted by

$$Proj[R; X_{i1} \times X_{i2} \times \cdots \times X_{ir}]$$

This is a fuzzy set P and its membership function is given by:

$$\mu_P(x_{i1}, x_{i2}, \cdots, x_{ir}) = \sup_{x_{j1}, x_{j2}, \cdots, x_{jm}} \{\mu_R(x_1, x_2, \cdots, x_n)\} \qquad (24)$$

The rationale for using the supremum operation on the membership function of R should be clear in view of the fact that we have a many-to-one mapping from an n-dimensional space to an r-dimensional space, with $r < n$.

As an example, consider the fuzzy set R shown in Figure 3(b) or the Cartesian product $A_1 \times A_2$ of two fuzzy sets (Figure 4(b)), both of which are fuzzy relations in the two-dimensional space $X_1 \times X_2$. For instance, the projection of R on X_1 has a membership function which is exactly the projection of $\mu_R(x_1, x_2)$ on the $\mu - x_1$ plane.

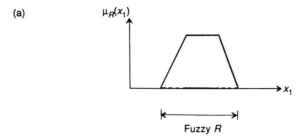

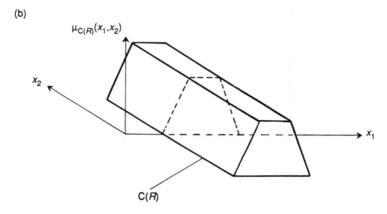

Figure 6 The cylindrical extension. (a) A fuzzy relation (set) and (b) its cylindrical extension.

Cylindrical Extension

Consider the Cartesian product space $X_1 \times X_2 \times \cdots \times X_n$, and suppose that the n indices are arranged as follows:

$$\{1, 2, \cdots, n\} \to \{i_1, i_2, \cdots, i_r, j_1, j_2, \cdots, j_m\}$$

Again, note that $r + m = n$ and that i and j denote the newly ordered set of n indices. Now consider a fuzzy relation R in the subspace $X_{i_1} \times X_{i_2} \times \cdots \times X_{i_r}$. Its cylindrical extension, denoted by $C(R)$, is given by

$$C(R) = \int_{X_1 \times X_2 \times \cdots \times X_n} \frac{\mu_R(x_{i_1}, x_{i_2}, \cdots, x_{i_r})}{x_1, x_2, \cdots, x_n} \tag{25}$$

Note that a cylindrical extension is a fuzzy set in the n-dimensional space and is the converse of projection. An example is given in Figure 6. Here, a fuzzy set R in the universe X_1 has been cylindrically extended to a fuzzy set in the Cartesian space $X_1 \times X_2$.

Join

Consider a fuzzy relation R in the subspace $X_1 \times X_2 \times \cdots \times X_r$ and a second fuzzy relation S in the subspace $X_m \times X_{m+1} \times \cdots \times X_n$ such that $m < r + 2$. Note that the union of these two subspaces gives the space $X_1 \times X_2 \times \cdots \times X_n$. The *join* of the fuzzy sets R and S is a fuzzy set in the $X_1 \times X_2 \times \cdots \times X_n$ space, and is given by the intersection of their cylindrical extensions; thus:

$$Join\,(R,S) = C(R) \wedge C(S) \text{ in } X_1 \times X_2 \times \cdots \times X_n \quad (26)$$

Its membership function is given by

$$\mu_{Join}(x_1, x_2, \cdots, x_n) = min[\mu_{C(R)}(x_1, x_2, \cdots, x_n), \mu_{C(S)}(x_1, x_2, \cdots, x_n)] \quad (27)$$

Note that *min* is applied here because the intersection of two fuzzy sets is considered.

Composition

Consider a fuzzy relation (fuzzy set) R in the subspace $X_1 \times X_2 \times \cdots \times X_r$ and a second fuzzy relation (fuzzy set) S in the subspace $X_m \times X_{m+1} \times \cdots \times X_n$ such that $m < r+1$. Note that unlike the previous case of join, the two subspaces are never disjoint, and hence their intersection is never null. As before, however, the union of the two subspaces gives $X_1 \times X_2 \times \cdots \times X_n$. The composition of R and S is denoted by $R \circ S$ and is given by:

$$R \circ S = Proj[Join(R,S); X_1, \cdots X_{m-1}, X_{r+1}, \cdots, X_n] \quad (28)$$

Here, we take the join of the two sets, as given by Equation (26), and then project the resulting fuzzy set onto the subspace formed by the disjoint parts of the two subspaces in which the fuzzy sets R and S are defined. The membership function of the resulting fuzzy set is obtained from the membership functions of R and S, while noting that *min* applies for *join* and *supremum* applies for *projection*. Specifically,

$$\mu_{(R \circ S)} = \sup_{X_m, \cdots, X_r} \{min(\mu_R, \mu_S)\} \quad (29)$$

Composition can be interpreted as a matching of two fuzzy sets, and making an inference according to the result. Specifically, the two sets are combined (join) and matched over their common subspace (supremum). The result is then projected over to the inference subspace, giving a fuzzy set defined over the disjoint portions of the two subspaces (projection). This process of matching is quite analogous to matching data with the condition part of a rule in rule-based control. In this regard composition plays a crucial role in fuzzy inference and control. Note that composition is a commutative operation: $R \circ S = S \circ R$.

Example

As a simple example, consider a fuzzy relation R defined in the $X \times Y$ space and another fuzzy relation S defined in the $Y \times Z$ space. The composition of these two fuzzy sets is given by

$$\mu_{R \circ S}(x, z) = \sup_{y \in Y} \{min(\mu_R(x, y), \mu_S(y, z))\} \quad (30)$$

Compositional Rule of Inference

In knowledge-based control systems control knowledge is often expressed as rules of the form

"**IF** output Y_1 is y_1 **THEN IF** output Y_2 is y_2 **THEN** control C is c"

In fuzzy control rules of this type are linguistic statements of expert knowledge in which y_1, y_2, and c are fuzzy quantities (e.g., small negative, fast, large positive). These rules are fuzzy relations that employ the fuzzy implication (IF-THEN). The collective set of rules forms the knowledge base in fuzzy control. Let us denote the fuzzy relation formed by this collection of rules as the fuzzy set R. In fuzzy control rules in R are first matched with available data (context). Next, a matched rule is fired, thereby providing a control action. Usually the context would be the measured outputs of the process, and these are crisp quantities, and the control action that drives the process is a crisp quantity as well. However, for general consideration suppose that the data (context) are denoted by a fuzzy set D and the control action is denoted by a fuzzy set C. The *compositional rule of inference* states that:

$$C = D \circ R \quad (31)$$

By using Equation (31) we can determine the membership function of the control action by incorporating the knowledge of the membership functions of data and rule base. Specifically, we have:

$$\mu_c = \sup_Y min(\mu_D, \mu_R) \quad (32)$$

This result follows directly from Equation (29). Note that Y denotes the space in which the data D are defined, and it is a subspace of the space in which the rule base R is defined. Furthermore, because R consists of fuzzy implications, its membership function can be formed by the constituent membership functions, using the *min* operation [see Equation (8)]. This method of obtaining R is analogous to model identification in conventional "hard" (crisp) control.

Example

Suppose that a fuzzy set A represents the output of a process and that it belongs to a discrete and finite universe Y of cardinality (number of elements) 5. Furthermore, suppose that a fuzzy set C represents the control input to the process and that it belongs to a discrete and finite universe Z of cardinality 4. It is given that:

$$A = 0.2/y_2 + 1.0/y_3 + 0.8/y_4 + 0.1/y_5$$
$$C = 0.1/z_1 + 0.7/z_2 + 1.0/z_3 + 0.4/z_4$$

A fuzzy relation R is defined by the fuzzy implication $A \rightarrow C$. The membership function of R is obtained using Equation (8); thus:

$$\mu_R(y_i, z_j) = \begin{bmatrix} 0 & 0 & 0 & 0 \\ 0.1 & 0.2 & 0.2 & 0.2 \\ 0.1 & 0.7 & 1.0 & 0.4 \\ 0.1 & 0.7 & 0.8 & 0.4 \\ 0.1 & 0.1 & 0.1 & 0.1 \end{bmatrix}$$

which is defined in the two-dimensional space $X \times Y$. This matrix represents the rule base in fuzzy control. Specifically, the information carried by the fuzzy rules $A \rightarrow C$ has been reduced to a relation R in the form of a matrix. Note that R is not a decision table for fuzzy control, even though a decision table can be derived using R by successively applying the compositional rule of inference to R for expected process responses.

Now suppose that a process measurement y_0 is made and that the measurement is closest to the element y_4 in Y. This is a crisp set and may be represented by a fuzzy singleton A_0 with a membership function:

$$\mu_{A0}(y_i) = [0, 0, 0, 0.8, 0]$$

60 INTELLIGENT CONTROL: FUZZY LOGIC APPLICATIONS

The membership function of the corresponding fuzzy control inference C' is obtained using the compositional rule of inference (Equation (32)). Specifically, we have:

$$\mu_{C'}(z_i) = \underset{\text{row}}{\max} \underset{\text{column}}{\min} \left\{ \begin{bmatrix} 0 \\ 0 \\ 0 \\ 0.8 \\ 0 \end{bmatrix}, \begin{bmatrix} 0 & 0 & 0 & 0 \\ 0.1 & 0.2 & 0.2 & 0.2 \\ 0.1 & 0.7 & 1.0 & 0.4 \\ 0.1 & 0.7 & 0.8 & 0.4 \\ 0.1 & 0.1 & 0.1 & 0.1 \end{bmatrix} \right\}$$

$$= \underset{\text{row}}{\max} \begin{bmatrix} 0 & 0 & 0 & 0 \\ 0 & 0 & 0 & 0 \\ 0 & 0 & 0 & 0 \\ 0.1 & 0.7 & 0.8 & 0.4 \\ 0 & 0 & 0 & 0 \end{bmatrix}$$

$$= [0.1, 0.7, 0.8, 0.4]$$

Hence, the fuzzy inference is given by

$$\mu_{C'}(z_i) = [0.1, 0.7, 0.8, 0.4]$$

This is simply the fourth row of μ_R. This fuzzy inference must be defuzzified (i.e., made crisp) for use in process control, e.g., using the *centroid method*, as discussed in Chapter 4.

■ MEMBERSHIP FUNCTION ESTIMATION

Establishment of the membership functions of input and output variables is an important first step of fuzzy control. The simplest and most common way of determining a membership function is to first decide on a discrete universe of discourse for the particular fuzzy set and then guess a grade of membership for each value in this universe. A triangular function peaked at the most representative value of the fuzzy quantity is commonly used for this purpose. More formal and systematic methods are available for estimating membership functions of fuzzy quantities. Five such methods are outlined in this section.

Averaged Guess Method

A distance function $d(x)$ is estimated using common sense or available information. This function is a measure of the distance of an element x from the fuzzy set A. If a value of x is known to be definitely an element of A, then the distance is taken to be zero. If a value is definitely not an element of A, then its distance is taken as some upper-bound value ($sup\ d$). The membership function is given by the relation:

$$\mu_A(x) = 1 - \frac{d(x)}{sup\ d} \tag{33}$$

An Intuitive Relation

It is intuitively clear that the rate of change of a membership function with respect to an element value should increase with both the strength of belief and the strength of disbelief that the value belongs to the set. Analytically we can express this by:

$$\frac{d\mu_A(x)}{dx} = k\mu_A(x)(1 - \mu_A(x)) \tag{34}$$

Direct integration of this relation gives:

$$\mu_A(x) = \frac{1}{1 + \exp(a - bx)} \tag{35}$$

The parameters a and b must be determined using additional knowledge of the problem.

Use of a Binary Polling

Each member of a group of experts is asked the question whether a specified value x is an element of the fuzzy set, allowing only yes or no answers. This polling is repeated for all the values of the (discrete) universe of discourse. Then, the membership function is estimated as

$$\mu_A(x) = \frac{\text{Number of yes answers}}{\text{Total number of answers}} \tag{36}$$

Relative Preference Method

Consider a fuzzy set

$$A = \{x_i : \mu_A(x_i)\} \tag{37}$$

in the discrete and finite universe X of cardinality n. The relative preference of x_i over x_j for membership of A is denoted by p_{ij}. A reasonable measure for this quantity is given by

$$p_{ij} = w_i / w_j \qquad (38)$$

in which

$$w_i = \mu_A(x_i) \qquad (39)$$

The matrix **P**, whose elements are given by Equation (38), is said to be consistent. Note that $p_{ii} = 1$ and that $p_{ji} = 1/p_{ij}$. This matrix has the following properties:

1. All eigenvalues of **P** are zero except one, which is n.
2. The eigenvector corresponding to this maximum eigenvalue (n) is the membership function vector having the discrete grades of membership w_i as elements.

It follows that the membership function of a fuzzy set can be determined by first estimating a relative preference matrix **P** and then computing the eigenvector corresponding to its maximum eigenvalue. A measure of consistency of this estimate would be λ_{max} which is the maximum eigenvalue of **P**. If this value is close to unity, the estimate is considered accurate.

■ SUMMARY

Conventional, two-state (or bivalent) logic cannot handle some important real-life situations. In particular, qualitative descriptors such as "fast" or "small" do not always possess crisp truth values (true or false). In other words, these are fuzzy terms. A more general type of logic that is applicable in such situations is *fuzzy logic*. Its mathematics is based on *fuzzy sets*.

A crisp set has its elements completely within the set, but the elements of a fuzzy set have *membership grades* associated with them. A membership grade can take a value in the interval [0, 1] and represents the degree of belongingness of an element to its fuzzy set. The *extension principle* is used to map one fuzzy set onto another fuzzy set through a crisp function.

Fuzzy logic is used to represent and process knowledge in fuzzy decision making systems. Specifically, fuzzy-logic operations such as AND, OR, NOT, and IF-THEN are employed and typically *sup* and *min* operations of membership functions are used. Compositional rule of inference is used to make decisions in a fuzzy system. Here, two fuzzy sets are matched and then projected onto the disjoint subspace within the product space of the two sets. These concepts are valuable in fuzzy-logic control.

PROBLEMS

1. Suppose that A denotes a set, X denotes the universal set, and ϕ denotes the null set.
   The law of contradiction
   $$A \wedge \overline{A} = \phi$$
   and the *law of excluded middle*
   $$A \vee \overline{A} = X$$
   are satisfied by crisp sets and hence agreeable to bivalent logic. Show that these two laws are not satisfied by fuzzy sets and hence violated in fuzzy logic. Give an example to illustrate each of these two cases.

2. Consider the laws of bivalent (crisp) logic as given in Table 2 of Chapter 2. Which of these laws are violated by fuzzy logic? In particular, show that De Morgan's laws are satisfied.

3. Two fuzzy sets A and B are represented by the following two membership functions:
   $$\mu_A(x) = max\left(0, \frac{x-3}{7}\right) \text{ for } x \leq 10$$
   $$= max\left(0, \frac{17-x}{7}\right) \text{ for } x > 10$$

   $$\mu_B(x) = max\left(0, \frac{x-8}{2}\right) \text{ for } x \leq 10$$
   $$= max\left(0, \frac{12-x}{2}\right) \text{ for } x > 10$$

   (a) Sketch these membership functions.
   (b) What do A and B approximately represent?
   (c) Which one of the two sets is fuzzier?

4. Consider a fuzzy set A in the universe $\Re$ (i.e., the real line) whose membership function is given by

$$\mu_A(x) = 1 - |x - 2| \text{ for } |x - 2| \leq 1$$
$$= 0 \text{ otherwise}$$

(a) Sketch the membership function.
(b) What is the support set of A?
(c) What is the α-cut of A for $\alpha = 0.5$?

5. A fuzzy set A is defined in the universe $X = [0, 8]$ and its membership function is given by

$$\mu_A(x) = \frac{x+1}{x+2}$$

Determine the fuzzy set B that is obtained through the crisp relation $y = x + 1$. What is the corresponding universe?

6. Consider the knowledge base represented by the set of fuzzy rules $Y \rightarrow U$ with

$$Y = \frac{0.2}{y_1} + \frac{0.6}{y_2} + \frac{1.0}{y_3} + \frac{0.6}{y_4} + \frac{0.2}{y_5}$$

and

$$U = \frac{0.3}{u_1} + \frac{1.0}{u_2} + \frac{0.3}{u_3}$$

Determine the membership function matrix R of this knowledge base.

Suppose that three fuzzy observations Y' have been made whose membership functions are
(a) $\mu_{Y'}(y_i) = [0.4\ 1.0\ 0.4\ 0\ 0]$
(b) $\mu_{Y'}(y_i) = [0.4\ 0.8\ 0.4\ 0\ 0]$
(c) $\mu_{Y'}(y_i) = [0.4\ 0.5\ 0.4\ 0\ 0]$

Determine the corresponding fuzzy inference U' in each case.

7. The *transitivity property* of conventional (crisp) sets states: If $A \subset B$ and $B \subset C$, then $A \subset C$. Is this property satisfied by fuzzy sets? Explain.

8. The *involution property* of fuzzy sets states that $\overline{\overline{A}} = A$ (i.e., NOT (NOT A) = A). Is this property satisfied by fuzzy sets? Explain.

9. The chronology of *Artificial Intelligence* may be summarized as follows:

    The first technical paper on the subject appeared in 1950. LISP programming language and the first expert system were developed in the 1960s. The first paper on "Fuzzy Sets" was published by Zadeh in 1964. Systems with natural language capabilities appeared in the 1970s. Many hardware tools, development systems, and expert systems were commercially available, and proliferation of fuzzy-logic applications took place in the 1980s. Integration of various AI techniques, extensive availability of PC-based systems, and sophisticated user interfaces are noted in the 1990s. In your opinion why did it take about two decades for us to accept the versatility to *Fuzzy Logic* in AI applications?

10. The characteristic function χ_A of a crisp set A is analogous to the membership function of a fuzzy set, and is defined as follows:

    $$\chi_A(x) = 1 \quad \text{if } x \in A$$
    $$= 0 \quad \text{otherwise}$$

    Show that

    $$\chi_{A'} = 1 - \chi_A$$
    $$\chi_{A \vee B} = max(\chi_A, \chi_B)$$
    $$\chi_{A \wedge B} = min(\chi_A, \chi_B)$$
    $$\chi_{A \to B}(x, y) = min[1, \{1 - \chi_A(x) + \chi_B(y)\}]$$

    where A and B are defined in the same universe X, except in the last case (implication) where A and B may be defined in two different universes X and Y.

    What are the implications of these results?

11. Show that the *commutative* and *associative* properties and *boundary conditions*, as listed in Table A.1 of Appendix A, are satisfied for "*min*" and "*max*" norms. Also show that De Morgan's law is satisfied by the two norms. This verifies that "*min*" is a T-norm and "*max*" is an S-norm.

12. For two membership functions μ_A and μ_B show that

$$\mu_A \cdot \mu_B \leq min(\mu_A, \mu_B)$$

This indicates that the "*max-dot*" inference tends to provide fuzzier inferences (more conservative) than the "*max-min*" inference.

13. If $\mu_A < 0.5$, show that a set A^* that is less fuzzy than A satisfies $\mu_{A^*} < \mu_A < 0.5$.

Similarly, if $\mu_A > 0.5$, a set A^* that is less fuzzy than A satisfies

$$\mu_{A^*} > \mu_A > 0.5$$

14. The degree of containment $c_{A,B}$ of a fuzzy set A in another fuzzy set B, both of which being defined in the same universe, is given by

$$c_{A,B} = 1 \text{ if } \mu_A \leq \mu_B$$
$$= \mu_B \text{ if } \mu_A > \mu_B$$

What does this parameter represent?

15. Consider the S-norm (or T-conorm) given by $x + y - xy$ with $0 < x < 1$ and $0 < y < 1$. Show that

$$x + y - xy > max(x, y)$$

Where $max(x, y)$ is another S-norm (See Table A.1 of Appendix A).

16. For fuzzy information $\mu_X(x)$ and a fuzzy rule base $\mu_R(x, y)$, the compositional rule of inference provides the inference

$$X \circ R = \sup_{x} min \, [\mu_X(x), \mu_R(x, y)]$$

Now consider the special case where the rule base is actually a crisp relation $y = f(x)$. Specifically,

$$\mu_R = 1 \quad \text{if } y = f(x)$$
$$= 0 \quad \text{otherwise}$$

Show that the extension principle is obtained in this case.

17. For a given fuzzy relation (rule base) $R(x, y)$, its eigen-fuzzy sets are defined as the family of fuzzy sets $\underline{E}$ such that

$$\underline{E} = \{E | E \circ R = E\}$$

where "∘" denotes the composition operation. Note here the analogy of eigenvectors of a matrix. Also, note that

$$R : X \times X \to [0,1]$$

Suggest an iterative method to obtain the greatest eigen-fuzzy set of R.

18. Show that $max[0, x + y - 1]$ is a T-norm, as given in Table A.1 of Appendix A. *Hint*: Show that the non-decreasing, commutative, and associative properties and the boundary conditions are satisfied.

19. Bayes' relation (or Bayes' theorem) is commonly used in probabilistic decision making (or, knowledge processing). This may be interpreted as a classification problem. Suppose that an observation y is made, and it may belong to one of several classes c_i. The Bayes' relation states

$$P(c_i|y) = \frac{P(y|c_i) \bullet P(c_i)}{P(y)} \qquad (1)$$

where

$P(c_i|y)$ = given that the observation is y, the probability that it belongs to class c_i (This is the *a posteriori conditional probability*);

$P(y|c_i)$ = given that the observation belongs to the class c_i, the probability that the observation is y. (This is the *class conditional probability*);

$P(c_i)$ = the probability that a particular observation belongs to class c_i, without knowing the observation itself. (This is the *a priori probability*);

$P(y)$ = the probability that the observation is y, without any knowledge of its class.

In the Bayes' decision making approach, for a given observation y, *a posteriori* probabilities $P(c_i|y)$ are computed for all possible classes ($i = 1, 2, .., n$), using Equation (1). The class that corresponds to the highest of these *a posteriori* probability values is chosen as the class of y, thereby solving the classification problem. The remaining $n - 1$ *a posteriori* probabilities represent the error in this decision. Discuss how knowledge may be represented in a probabilistic approach of decision making (of which Bayes' approach is an example).

REFERENCES

Dubois, D. and Prade, H. (1980). *Fuzzy Sets and Systems*, Academic Press, Orlando, FL.

Klir, G.J. and Folger, T.A. (1988). *Fuzzy Sets, Uncertainty, and Information*, Prentice-Hall, Englewood Cliffs, NJ.

Kosko, B. (1992). *Neural Networks and Fuzzy Systems*, Prentice-Hall, Englewood Cliffs, NJ.

Pedrycz, W. (1989). *Fuzzy Control and Fuzzy Systems*, Research Studies Press, Ltd., Somerset, England.

Zadeh, L.A. (1973). Outline of a New Approach to the Analysis of Complex Systems and Decision Processes, *IEEE Trans. Systems, Man and Cybernetics*, Vol. 3, No. 1, pp. 28–44.

Zadeh, L.A. (1979). A Theory of Approximate Reasoning, in *Machine Intelligence*, J.E. Hayes, D. Michie, and L.I. Mikulich (Eds.), Vol. 9, pp. 149–194, Elsevier, New York.

4 FUZZY LOGIC CONTROL

■ INTRODUCTION

It is common knowledge in process control practice that when an experienced control engineer is included as an advisor to a control loop, significant improvements in the system performance are usually possible. Generally, adjusting operating variables and control settings, tuning, and other control actions carried out by a human operator are implemented, not through hard (crisp) algorithms, but rather qualitatively using linguistic rules which are based on common sense, heuristics, knowledge, and experience. For example, an expert might teach a process operator the necessary control actions using a set of protocols containing linguistic fuzzy terms such as "fast", "small", and "accurate". A practical difficulty invariably arises because except in very low bandwidth processes, human actions are not fast enough for this approach to be feasible. Furthermore, it is not economical to dedicate a human expert to every process, as human experts are scarce. In fuzzy logic control the task of generating control decisions is computer automated, thereby alleviating the problems of control speed and facilitating simultaneous monitoring and control of many variables. There, linguistic descriptions of human expertise in controlling a process are represented as fuzzy rules or relations, and this knowledge base is used, in conjunction with some knowledge of the state of the process (e.g., of measured response variables), by an inference mechanism to determine control actions at a sufficiently fast rate.

Fuzzy logic control does not require a conventional model of the process, whereas most conventional control techniques (i.e., model-based control) require either an analytical model or an experimental model, as discussed in Chapter 1. In view of this, fuzzy logic control is particularly suitable for complex and ill-defined processes in which analytical modeling is difficult due to the fact that the process is not completely known, and experimental model identification is not feasible because the required inputs and outputs of the process may not be measurable. Being a knowledge-based control approach it has advantages when considerable experience is available regarding the behavior of the process and when this knowledge may be expressed as a set of rules containing fuzzy quantities.

Fuzzy control has been applied in a number of processes ranging from ship autopilots, subway cars, helicopters, and pilot-scale steam engines to cement kilns, robotic manipulators, fish processing machines, and household appliances. One obvious drawback in many such applications is that fuzzy control laws are implemented at the lowest level — within a servo or direct-digital-control (DDC) loop (de Silva and MacFarlane, 1989b), generating control signals for process actuation directly through fuzzy inference. In high-bandwidth processes this form of fuzzy control implementation would require very fast and accurate control in the presence of strong nonlinearities and dynamic coupling. Another drawback of this direct implementation of fuzzy control is that the control signals are derived from inferences that are fuzzy, thereby directly introducing errors into the necessary crisp control signals. A third argument against the conventional, low-level implementation of fuzzy control is that in a high-speed process, human experience is gained not through manual, on-line generation of control signals in response to process output, but typically through performing parameter adjustments and tuning (manual adaptive control) operations. Hence, it can be argued that, in this case, the protocols for generating control signals are established not through direct experience but by some form of correlation. Of course, it is possible to manually examine input-output data recorded from a high-speed process. However, in such off-line studies, on-line human interaction is not involved, and again the experience gained would be indirect and somewhat artificial. It follows that it is more appropriate to use fuzzy logic for tuning a controller and for similar high-level tasks rather than in direct control (de Silva and MacFarlane, 1988; 1989a).

This chapter presents the basics of fuzzy logic control, as well as the method of generating a membership function matrix for a fuzzy rule base. Next, the process of making control decisions by applying the *compositional rule of inference* (see Chapter 3) is explained. Typically the monitored process variables are crisp and they must be *fuzzified* for the purpose of decision making. Also, the control decisions generated in this manner are fuzzy quantities, and they must be *defuzzified* prior to using them in the physical action of control. These methods are described. We make some attempt to differentiate between high-level control and low-level direct control, and it is hinted that fuzzy logic is more suited for the former. Examples are given to illustrate the various steps of fuzzy logic control.

The starting point of conventional fuzzy control is the development of a rule base using linguistic descriptions of control protocols, e.g., of a human expert. This step is analogous to the development of a "hard" (crisp) control algorithm and the identification of parameter values for the algorithm in a conventional control approach. The rule base consists of a set of IF-THEN rules relating fuzzy quantities which represent process response (outputs) and control inputs. During the control action, process measurements (context) are matched with rules in the rule base, using the compositional rule of inference (Zadeh, 1975) to generate fuzzy control inferences. This inference procedure is clearly analogous to feedback control in a hard control scheme.

■ BASICS OF FUZZY CONTROL

A rule in a fuzzy knowledge base is generally a relation of the form:

$$\text{IF } A_i \text{ THEN IF } B_i \text{ THEN } C_i \tag{1}$$

where, A_i and B_i are fuzzy quantities representing process measurements (e.g., process error and change in error) and C_i is a fuzzy quantity representing a control signal (e.g., change in process input). Because these fuzzy sets are related through IF-THEN implications and because an implication operation for two fuzzy sets can be interpreted as a "minimum operation" on the corresponding membership functions, the membership function of this fuzzy relation may be expressed as

$$\mu_{Ri}(a,b,c) = min[\mu_{Ai}(a), \mu_{Bi}(b), \mu_{Ci}(c)] \tag{2}$$

The individual rules in the rule base are joined through ELSE connectives which are *unions* (OR connectives). Hence, the overall membership function for the complete rule base (relation R) is obtained using *maximum* operations on membership functions of the individual rules; thus

$$\mu_R(a,b,c) = \max_i \mu_{Ri}(a,b,c) \tag{3}$$

In this manner membership function of the entire rule base can be "identified" using membership functions of the response variables and control inputs. Note that a fuzzy knowledge base is a multidimensional array [a three-dimensional array in the case of Equation (3)] of membership function values. This array corresponds to a fuzzy control algorithm.

Once a fuzzy control algorithm of the form given by Equation (3) is obtained, we need a procedure to infer control actions using process measurements during control. Specifically, suppose that fuzzy process measurements A' and B' are available. The corresponding control inference C' is obtained using the compositional rule of inference (i.e., inference using the composition relation). As discussed in Chapter 3, the applicable relation is (Dubois and Prade, 1980)

$$\mu_{C'}(c) = \sup_{a,b} min[\mu_{A'}(a), \mu_{B'}(b), \mu_R(a,b,c)] \qquad (4)$$

Note that in fuzzy inference the fuzzy sets A' and B' are jointly matched with the fuzzy relation R. This is an intersection which corresponds to an AND operation, and hence *min* operation applies for the membership functions. For a given value of control action c, the resulting fuzzy sets are then mapped (projected) from a three-dimensional space $X \times Y \times Z$ onto a one-dimensional space Z of control actions. This mapping corresponds to a set of OR connectives, and hence *sup* operation applies to the membership function values, as expressed in Equation (4).

Actual process measurements are crisp. Hence, they have to be fuzzified in order to apply the compositional rule of inference. This is done by reading off the membership grades corresponding to the measurement using the membership functions of the measurement. Conversely, the control action must be a crisp value. Hence, each control inference C' must be defuzzified so that it can be used to control the process. Several methods are available to accomplish this. In the *mean of maxima* method the control element corresponding to the maximum grade of membership is used as the control action. If more than one element with maximum membership value is present, the mean of these values is used. In the *center of gravity (or centroid) method* the centroid of the membership function of control decision is used as the value of crisp control action. This weighted control action is known to provide a sluggish, yet more robust control. These methods are discussed further later in the chapter. Schematic representation of a fuzzy controller of the type described here is shown in Figure 1. In this diagram application of the compositional rule of inference has been interpreted as a rule-matching procedure.

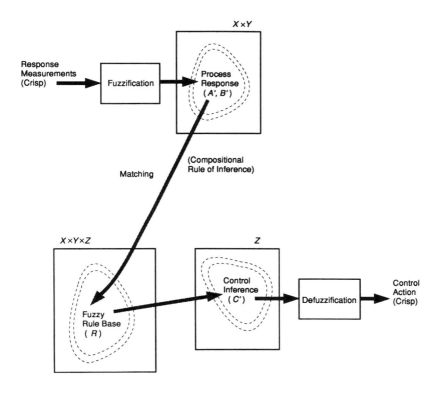

Figure 1 A fuzzy logic controller.

There are several practical considerations of fuzzy control that were not addressed in the above discussion. Because representation of the rule base R by an analytical function is unrealistic and infeasible in general, it is customary to assume that the fuzzy sets involved in R have discrete and finite universes (or at least discrete and finite support sets). As a result, process response measurements must be quantized. Hence, at the outset, a decision must be made as to the element resolution (quantization error) of each universe. This resolution governs the cardinality of the sets and in turn the size of the multidimensional membership function array of a fuzzy rule base. It follows that computational effort, memory and storage requirements, and accuracy are directly affected by quantization resolution. Because process measurements are crisp, one method to reduce real-time computational overhead is to precompute a decision table relating quantized measurements to crisp control actions.

The main disadvantage of this approach is that it does not allow for modifications (e.g., rule changes and quantization resolution adjustments) during operation. Another practical consideration is the selection of a proper sampling period in view of the fact that process responses are generally analog signals. Factors such as process characteristics, required control bandwidth, and the processing time needed for one control cycle, must be taken into account in choosing a sampling period. Scaling or gain selection for various signals in a fuzzy logic control system is another important consideration. For reasons of processing efficiency, it is customary to scale the process variables and control signals in a fuzzy control algorithm. Furthermore, adjustable gains can be cascaded with these system variables so that they may serve as tuning parameters for the controller. A proper tuning algorithm would be needed, however. A related consideration is real-time modification of a fuzzy rule base. Specifically, rules may be added, deleted, or modified on the basis of some *self-organization* scheme (Procyk and Mamdani, 1979). For example, using a model for the process and making assumptions such as input-output monotonicity, it is possible during control to trace and tag the rules in the rule base that need attention. The control-decision table can be modified accordingly.

Steps of Fuzzy Logic Control

The main steps of fuzzy logic control, as described above, can be summarized in the following manner. A fuzzy control algorithm must be developed first (off-line), according to the following four steps:

1. Develop a set of linguistic control rules (protocols) that may contain fuzzy variables as conditions (process outputs) and actions (control inputs to the process).
2. Obtain a set of discrete membership functions for process output variables and control input variables.
3. Using fuzzy implication on each rule in step 1, and using step 2, obtain the multi-dimensional array R_i of membership values for that rule.
4. Combine the relations R_i using fuzzy connectives (AND, OR, NOT) to obtain the overall fuzzy rule base (relation R).

Then, control actions may be determined in real time as follows:

1. Fuzzify the measured process variables as fuzzy singletons (membership grade values that are read off the membership functions of these variables).

2. Match the fuzzy measurements obtained in step 1 with the membership array of the fuzzy rule base (obtained in step 4), using the compositional rule of inference.
3. Defuzzify the control inference obtained in step 2 (either the *mean of maxima method* or the *centroid method* may be used here).

These steps reflect the formal procedure in fuzzy control. There are several variations. For example, a much faster approach would be to develop a crisp decision table by combining the four steps of fuzzy algorithm development and the first two steps of control, and using this table in a table look-up mode to determine a crisp control action during operation.

■ DECISION MAKING WITH CRISP MEASUREMENTS

The compositional rule of inference as represented by Equation (4) assumes that the systems measurements used are themselves fuzzy. These measurements are represented by fuzzy labels (sets) A' and B'. This is true for "human-like", *high-level sensors* in which the sensor itself is treated as a fuzzy system. This is also true if the information from low-level crisp sensors are preprocessed by *intelligent preprocessors* or *intelligent filters* so as to generate fuzzy information for subsequent application of the compositional rule of inference as in Equation (4).

More common, at least in low-level direct control, is sensory information that is crisp. This information is used in Equation (4) without intelligent preprocessing. In this case the application of the compositional rule of inference becomes quite simple. To derive the result corresponding to Equation (4) for this case, we proceed as follows: suppose that the rule base is given by

$$\underset{i,j}{Else}(A_i, B_j) \to C_k \tag{5}$$

in which the possible combinations of fuzzy states (i,j) of A and B depend on the number of rules in the rule base, and the fuzzy state k of C is determined by the particular rule considered, and hence is dependent on (i,j).

Define a delta function as,

$$\begin{aligned}\delta(p - p_0) &= 1 \quad \text{when} \quad p = p_0 \\ &= 0 \text{ elsewhere}\end{aligned} \tag{6}$$

76 INTELLIGENT CONTROL: FUZZY LOGIC APPLICATIONS

Next suppose that the crisp measurements (a_0, b_0) are measured from the process outputs (a, b). These may be represented as

$$\mu_{A'}(a) = \delta(a - a_0); \quad \mu_{B'}(b) = \delta(b - b_0) \tag{7}$$

The corresponding control decision C' is determined by applying the compositional rule of interference [Equation (3.29)] as

$$\mu_{C'}(c) = \sup_{a,b} min[\delta(a - a_0), \delta(b - b_0), \max_{i,j} min(\mu_{A_i}(a), \mu_{B_j}(b), \mu_{C_k}(c))] \tag{8}$$

Note that the last term on the RHS of Equation (8) is the rule base $\mu_R(a, b, c)$ in the expanded form. The two "min" operations in this equation may be combined and also, in view of the nature of the delta functions [Equation (6)], they may be absorbed into the combined "min". This gives

$$\mu_{C'}(c) = \sup_{a,b} \max_{i,j} min[\mu_{A_i}(a) \cdot \delta(a - a_0), \mu_{B_j}(b) \cdot \delta(b - b_0), \mu_{C_k}(c)] \tag{9}$$

Finally, Equation (6) is used in (9) to obtain the result

$$\mu_{C'}(c) = \max_{i,j} min[\mu_{A_i}(a_0), \mu_{B_j}(b_0), \mu_{C_k}(c)] \tag{10}$$

This is the most common form of the compositional rule of inference that is employed in decision making for fuzzy logic control. The steps of applying Equation (10) are as follows:

1. Consider the first rule, $i = 1$. Initiate $\mu_{C_0}(c) = 0$.
2. For this rule (i) read the membership grades at points $a = a_0$ and $b = b_0$ in the process output membership functions $\mu_{A_i}(a)$ and $\mu_{B_i}(b)$.
3. Take the minimum of the two values of step 2. Clip the process output membership function $\mu_{C_i}(c)$ with this value.
4.

$$\mu_{C_i}(c) = \sup_c [\mu_{C_i}(c), \mu_{C_{i-1}}(c)]$$

5. If i = total number of rules, set $\mu_{C'}(c) = \mu_{C_i}(c)$ and stop. Otherwise, set $i = i + 1$ and go to step 2.

It should be clear that in making control inferences, Equation (10) is computationally far more attractive than Equation (4). In particular, in Equation (10) each rule is considered separately and only in terms of the membership grades of the constituent condition

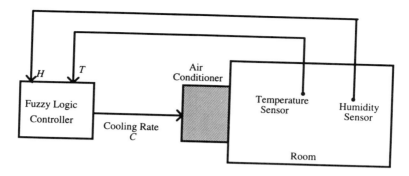

Figure 2 Comfort control system of a room.

variables at the measurement point. Equation (4) assumes that the multidimensional membership function of the entire rule base (R) is available which involves a large number of *min* and *max* operations. Then the compositional rule of inference is separately applied to this membership function, which requires another set of *min* and *sup* operations.

Example

Consider the room comfort control system schematically shown in Figure 2. The temperature (T) and humidity (H) are the process variables that are measured. These sensor signals are provided to the fuzzy logic controller which determines the cooling rate (C) that is generated by the air conditioning unit. The objective is to maintain a particular comfort level inside the room.

The fuzzy rule base of the comfort controller is shown in Figure 3. The temperature level can assume one of two fuzzy states (HG, LW), which denote high and low, with the corresponding membership functions. Similarly, the humidity level can assume two other fuzzy states (HG, LW) with associated membership functions. Note that the membership functions of T are quite different from those of H, even though the same nomenclature is used. There are four rules, as given in Figure 3.

Application of the compositional rule of inference is done here by using Equation (10). For example, suppose that the room temperature is 30°C and the relative humidity is 0.9. Lines are drawn at these points, as shown in Figure 3, to determine the corresponding membership grades for the fuzzy states in the four rules. In each rule the lower value of the two grades of process state is then used to clip (or modulate) the corresponding membership function of C (a *min* operation). The resulting "clipped" membership functions of C for all four rules are superimposed (a *max* operation) to obtain the control inference C', as shown. This result is a fuzzy set, and it must be defuzzified to obtain a crisp

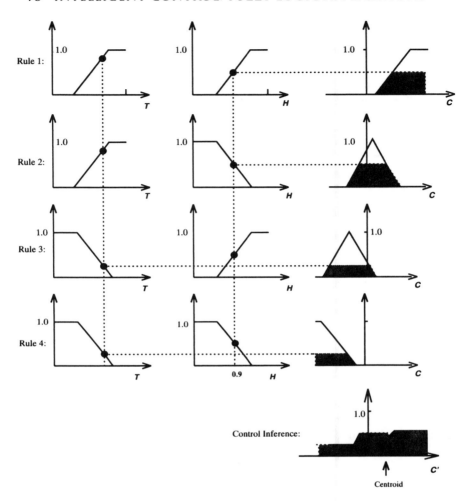

Figure 3 The fuzzy knowledge base of the comfort controller.

control action $\hat{C}$ for changing the cooling rate. The centroid method may be used, as explained in the next section.

■ DEFUZZIFICATION

The output decision of a fuzzy logic controller is a fuzzy value and is represented by a membership function. Because low-level control actions are typically crisp, the control inference must be *defuzzified* for actuation purposes. Several methods are available for defuzzification of a fuzzy control inference.

FUZZY LOGIC CONTROL

Centroid Method

Suppose that the membership function of a control inference is $\mu_C(c)$, with its *support set* given by

$$S = \{c \mid \mu_C(c) > 0\} \tag{11}$$

Then, the centroid method of defuzzification is expressed as

$$\hat{c} = \frac{\int_{c \in S} c \mu_C(c) dc}{\int_{c \in S} \mu_C(c) dc} \tag{12}$$

where $\hat{c}$ is the defuzzified control action.

Note that Equation (12) corresponds to the case in which the membership function is continuous. The discrete case is given by

$$\hat{c} = \frac{\sum_{c_i \in S} c_i \mu_C(c_i)}{\sum_{c_i \in S} \mu_C(c_i)} \tag{13}$$

Mean of Maxima Method

If the membership function of the control inference is *unimodal* (i.e., it has just one peak value), the control signal at the peak value is chosen as the defuzzified control action, in the mean of maxima method. Specifically

$$\hat{c} = c_{max} \text{ such that } \mu_C(c_{max}) = \max_{c \in S} \mu_C(c) \tag{14}$$

The result for the discrete case follows from this relation.

If the control membership function is *multimodal* (i.e., has more than one peak), the mean value of this set of peaks, weighted by the corresponding membership grades, is used as the defuzzified value. Hence, if we have

$$c_i \text{ such that } \mu_C(c_i) \triangleq \mu_i = \max_{c \in S} \mu_C(c) \quad i = 1, 2, \cdots, p$$

then

$$\hat{c} = \frac{\sum_{i=1}^{p} \mu_i c_i}{\sum_{i=1}^{p} \mu_i} \qquad (15)$$

Here, p is the total number of modes (peaks) in the membership function.

Threshold Methods

Sometimes it may be desirable to leave out the boundaries of the control inference membership function when computing a defuzzified value. In this manner, only the main core of the control inference is used, not excessively diluting or desensitising the defuzzification. Here, the set over which the integration, summation, or peak search is carried out is not the entire support set but an α-cut (see Chapter 3) of the control inference set. Specifically, we select

$$S_\alpha = \{c | \mu_c(c) \geq \alpha\} \qquad (16)$$

in which α is the threshold value. Then any one of Equations (12) through (15) may be used, with the support set S replaced by the α-cut S_α.

Comparison of the Defuzzification Methods

The *centroid method* uses the entire membership function of a control inference. Thus, the defuzzified value depends on both size and shape of the membership function. As a result it is a more complete (first moment) representation of the inference. However, due to inherent averaging over the support set, the control action is diluted and less sensitive to small variations. Accordingly, this represents a very robust procedure, that generates less oscillatory process responses.

The *mean of maxima method* depends only on the peak values of the control inference membership function. It is therefore independent of the membership function shape, and particularly its asymmetry. Consequently, it is a somewhat incomplete representation of the control inference. The resulting control actions tend to be more sensitive and sharper in this method as compared to those from the centroid method. This is so because a shift of the peak position will not change the centroid position by the same distance in general. As a result the approach is less robust.

Thresholding, using the α-cut given by Equation (16) for the region of integration, will make the centroid method more sensitive, less robust, and less dependent on the shape of the control membership function. In particular, in the limit of $\alpha = 1$, there is no difference

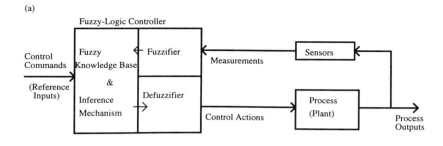

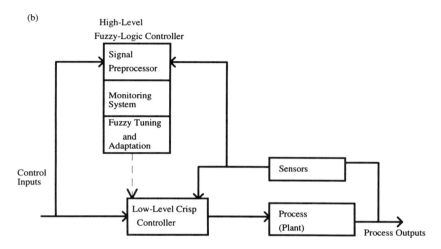

Figure 4 Two architectures of fuzzy logic control. (a) Low-level direct control. (b) High-level tuning/adaptive control.

between the centroid method and the mean of maxima method. Note that when the membership grades at the modal points of the control inference are large (close to 1.0), thresholding has no effect on the defuzzified value.

■ ARCHITECTURES OF FUZZY CONTROL

Two architectures of fuzzy logic control are common, and are shown in Figure 4. The conventional architecture, as shown in Figure 4(a), is by far the most common (Mamdani, 1977; Sugeno, 1985; Pedrycz, 1989). Here, the low-level direct control signals are generated by the fuzzy-logic controller. As pointed out in the beginning of this chapter (de Silva and MacFarlane, 1989a), this approach has some drawbacks with respect to speed, accuracy, and sensitivity and is not recommended unless some of the following conditions are satisfied:

1. Process time constants are relatively high (e.g., 1.0 s or more).
2. The required control bandwidth is sufficiently low (e.g., 1.0 Hz or less).
3. The process is complex and ill defined. Analytical modeling is difficult and experimental model identification is infeasible.
4. Sufficient past experience with low-level human-in-the-loop control is available and may be expressed as a set of linguistic protocols with fuzzy terms.

A more desirable architecture is the hierarchical structure (de Silva and MacFarlane, 1988), as shown in Figure 4(b), and further evolved in Chapters 5 through 8. The low-level controller is conventional and crisp, in the category outlined in Chapter 1. The fuzzy-logic controller operates at a higher level in this architecture, and typically performs tuning and adaptive control functions rather than in-loop direct control. Because high-level decision making requires high-level information, an important component in the control system is a signal preprocessor. The resulting (preprocessed) information is generally fuzzier and has a lower resolution, as discussed in Chapter 7. In this architecture a monitoring system uses high-level process information to evaluate the performance of the plant. On this basis, a fuzzy decision making unit makes adjustments to the control system. These adjustments may include on-line and off-line tuning of the controller parameters, on-line adaptation of the low-level controller, and self-organization and dynamic restructuring (de Silva, 1993b) of the control system.

Example of Fuzzy Tuner (Decision Table Approach)

This section outlines a fuzzy tuner for a proportional-integral-derivative (PID) servocontroller (de Silva and MacFarlane, 1988; 1989a). First, expert knowledge on tuning a PID servo is expressed as a set of linguistic statements. Such statements (rules) will necessarily contain fuzzy quantities. Next, membership functions are established for the fuzzy quantities which are present in the fuzzy rules. Each fuzzy rule is then expressed as a fuzzy relation, in tabular form, by incorporating the membership functions into it through fuzzy logic connectives (e.g., IF-THEN, OR). The resulting set of tables forms the rule base of the fuzzy tuner. Rule matching can be accomplished by on-line application of the compositional rule of inference to the fuzzy rule base if so desired, as explained previously. This on-line application of the compositional rule is generally time consuming, and hence it may considerably reduce the speed of servo tuning. Because the universe of discourse of the condition set is discrete and finite in the present application, and because the universe of the tuning action is also discrete and finite (both classes of universes

If the response oscillations are low then slightly decrease the proportional gain and slightly increase the derivative time constant; or if the response is moderately oscillatory, then moderately decrease the proportional gain and moderately increase the derivative time constant; or if the response is highly oscillatory, then make a large decrease in the proportional gain and a large increase in the derivative time constant; or if the response is extremely oscillatory, then make a large decrease in the proportional gain and a large increase in the derivative time constant and a slight decrease in the integral rate.

If the response is slow then increase the derivative time constant slightly and increase the proportional gain moderately; or if the speed of response is moderate then increase the derivative time constant slightly and increase the proportional gain slightly; or if the speed of response is high increase the derivative time constant slightly and decrease the proportional gain slightly; or if the speed of response is very high decrease the proportional gain moderately.

If the response diverges slowly then slightly decrease the proportional gain and slightly increase the derivative time constant; or if the response diverges moderately then slightly decrease the proportional gain and moderately increase the derivative time constant; or if the response diverges rapidly then slightly decrease the proportional gain and increase the derivative time constant by a large amount and slightly decrease the integral rate; or if the response diverges very rapidly then moderately decrease the proportional gain and increase the derivative time constant by a large amount and moderately decrease the integral rate.

If the offset is low then slightly increase the proportional gain, and slightly increase the integral rate; or if the offset is moderate then moderately increase the integral rate; or if the offset is high then increase the proportional gain slightly and increase the integral rate by a large value; or if the offset is very high then moderately increase the proportional gain and increase the integral rate by a large value.

Figure 5 A typical set of linguistic protocols for tuning a PID servo.

having low cardinality), computational efficiency can be significantly improved by applying the compositional rule of inference off line. This preprocessing will generate a *decision table* for fuzzy tuning of a servo controller. The steps of developing a fuzzy decision table for servo tuning are outlined below.

Consider a set of linguistic statements given in Figure 5. These linguistic rules reflect the actions of a human expert in tuning a PID servo by observing the error response of the servo. Of course, more rules can be added, and the resolution (the number of fuzzy states) can be increased to improve tuning accuracy, but for the purpose of the present demonstration, this rule set is adequate. We can express the fuzzy tuning rules in the condensed form shown in Figure 6. The notation used in the condensed form of the rule base is given below.

  OSC = Oscillations in the error response
  RSP = Speed of response (decay) of the error response
  DIV = Divergence (growth) of the error response

	if	OSC=LOW	then	DP=NL
			and	DD=PL
or	if	OSC=MOD	then	DP=NM
			and	DD=PM
or	if	OSC=HIG	then	DP=NH
			and	DD=PH
or	if	OSC=VHI	then	DP=NH
			and	DI=NL
			and	DD=PH
	if	RSP=LOW	then	DP=PM
			and	DD=PL
or	if	RSP=MOD	then	DP=PL
			and	DD=PL
or	if	RSP=HIG	then	DP=NL
			and	DD=PL
or	if	RSP=VHI	then	DP=NM
	if	DIV=LOW	then	DP=NL
			and	DD=PL
or	if	DIV=MOD	then	DP=NL
			and	DD=PM
or	if	DIV=HIG	then	DP=NL
			and	DI=NL
			and	DD=PH
or	if	DIV=VHI	then	DP=NM
			and	DI=NM
			and	DD=PH
	if	OFS=LOW	then	DP=PL
			and	DI=PL
or	if	OFS=MOD	then	DI=PM
or	if	OFS=HIG	then	DP=PL
			and	DI=PH
or	if	OFS=VHI	then	DP=PM
			and	DI=PH

Figure 6 Condensed form of tuning protocols.

```
OFS = Offset in the error response
DP  = Adjustment in the proportional gain
DI  = Adjustment in the integral rate parameter
DD  = Adjustment in the derivative time constant
NON = None
LOW = Low value
```

MOD = Moderate value
HIG = High value
VHI = Very high value
NH = Negative high value
NM = Negative moderate value
NL = Negative low value
NC = No change
PL = Positive low value
PM = Positive moderate value
PH = Positive high value

Next we must develop membership functions for the fuzzy conditions OSC, RSP, DIV, and OFS and for the fuzzy actions DP, DI, and DD. As described in Chapter 3, several methods for estimating such membership functions are available. For our purposes, an approximate set of membership functions would suffice, based on intuition. Generally, there is freedom to choose the resolution of each fuzzy quantity. In the present application, however, because the condition variables are constrained to a discrete and finite set, the resolution of the condition variables would be fixed. This restriction of resolution is not necessary for action variables. For example, even though the action variable DP can assume one of seven fuzzy values, the universe of one (e.g., NH), may contain more than seven elements. However, in the interest of consistency and simplicity, the cardinality of the universe of discourse of a fuzzy quantity is taken here to be equal to the number of fuzzy values which can be assumed by the particular attribute.

In applications of the present type actual numerical values given to the elements in a universe of a fuzzy quantity are of significance only in a relative sense, but an appropriate physical meaning should be attached to each value. For instance, in defining DP, the numerical value -3 is chosen to represent a negative high change in the proportional gain. This choice is compatible with the choice of -1 to represent a negative low change. Also, such numerical values may be scaled and converted as necessary to achieve physical and dimensional compatibility.

The chosen membership functions are shown in Tables 1 and 2. Note that a membership grade of unity is assigned as desired to the representative value of each fuzzy quantity. Fuzziness is introduced by assigning uniformly decreasing membership grades to the remaining element values.

The following shows the development of a fuzzy relation table for a group of fuzzy rules. For example, consider the rule

$$\text{If OSC = LOW then DP = NL} \qquad (17)$$

TABLE 1
MEMBERSHIP FUNCTIONS OF THE CONDITION VARIABLES

1. Oscillation

OSC	0	1	2	3	4
NON	1.0	0.8	0.6	0.4	0.2
LOW	0.8	1.0	0.8	0.6	0.4
MOD	0.6	0.8	1.0	0.8	0.6
HIG	0.4	0.6	0.8	1.0	0.8
VHI	0.2	0.4	0.6	0.8	1.0

2. Speed of Response

RSP	0	1	2	3	4
NON	1.0	0.8	0.6	0.4	0.2
LOW	0.8	1.0	0.8	0.6	0.4
MOD	0.6	0.8	1.0	0.8	0.6
HIG	0.4	0.6	0.8	1.0	0.8
VHI	0.2	0.4	0.6	0.8	1.0

3. Divergence

DIV	0	1	2	3	4
NON	1.0	0.8	0.6	0.4	0.2
LOW	0.8	1.0	0.8	0.6	0.4
MOD	0.6	0.8	1.0	0.8	0.6
HIG	0.4	0.6	0.8	1.0	0.8
VHI	0.2	0.4	0.6	0.8	1.0

4. Offset

OFS	0	1	2	3	4
NON	1.0	0.8	0.6	0.4	0.2
LOW	0.8	1.0	0.8	0.6	0.4
MOD	0.6	0.8	1.0	0.8	0.6
HIG	0.4	0.6	0.8	1.0	0.8
VHI	0.2	0.4	0.6	0.8	1.0

To construct its fuzzy relation table, we take the membership function of LOW in OSC from Table 1 and the membership function of NL in DP from Table 2. Next, in view of the fact that fuzzy implication is a *min* operation on membership grades, we form the Cartesian product space of these two membership function vectors and assign the lower value of each pair of membership grades to the corresponding location in the Cartesian space. This corresponds to the application of Equation (2). The other three relation tables representing the remaining three rules (see Figure 6) that connect OSC to DP are obtained in

TABLE 2
MEMBERSHIP FUNCTIONS OF THE ACTION VARIABLES

1. Proportional Gain

DP	-3	-2	-1	0	1	2	3
NH	1.0	0.6	0.2	0	0	0	0
NM	0.6	1.0	0.6	0.2	0	0	0
NL	0.2	0.6	1.0	0.6	0.2	0	0
NC	0	0.2	0.6	1.0	0.6	0.2	0
PL	0	0	0.2	0.6	1.0	0.6	0.2
PM	0	0	0	0.2	0.6	1.0	0.6
PH	0	0	0	0	0.2	0.6	1.0

2. Integral Rate

DI	-3	-2	-1	0	1	2	3
NH	1.0	0.6	0.2	0	0	0	0
NM	0.6	1.0	0.6	0.2	0	0	0
NL	0.2	0.6	1.0	0.6	0.2	0	0
NC	0	0.2	0.6	1.0	0.6	0.2	0
PL	0	0	0.2	0.6	1.0	0.6	0.2
PM	0	0	0	0.2	0.6	1.0	0.6
PH	0	0	0	0	0.2	0.6	1.0

3. Derivative Time Constant

DD	-3	-2	-1	0	1	2	3
NH	1.0	0.6	0.2	0	0	0	0
NM	0.6	1.0	0.6	0.2	0	0	0
NL	0.2	0.6	1.0	0.6	0.2	0	0
NC	0	0.2	0.6	1.0	0.6	0.2	0
PL	0	0	0.2	0.6	1.0	0.6	0.2
PM	0	0	0	0.2	0.6	1.0	0.6
PH	0	0	0	0	0.2	0.6	1.0

a similar fashion. The corresponding four rules are connected by fuzzy OR connectives. Hence, the composite relation table is obtained by combining the individual tables through a *max* operation. This corresponds to the application of Equation (3). The composite relation tables obtained in this manner are given in Table 3.

The final step in the development of the fuzzy tuner is the establishment of the decision table. Each inference of servo response (condition) must be matched with the rule base obtained in Table 3. This is accomplished by applying the *compositional rule of inference,* as given by Equation (4). Recall that this is a *sup of min operation.* Specifically, we compare the membership function vector of a response

TABLE 3
COMPOSITE FUZZY-RELATION TABLES

(a) OSC to DP

		\ DP						
		-3	-2	-1	0	1	2	3
	0	0.6	0.6	0.8	0.6	0.2	0	0
	1	0.6	0.8	1.0	0.6	0.2	0	0
OSC	2	0.8	1.0	0.8	0.6	0.2	0	0
	3	1.0	0.8	0.6	0.6	0.2	0	0
	4	1.0	0.6	0.6	0.4	0.2	0	0

(b) OSC to DI

		\ DI						
		-3	-2	-1	0	1	2	3
	0	0.2	0.2	0.2	0.2	0.2	0	0
	1	0.2	0.4	0.4	0.4	0.2	0	0
OSC	2	0.2	0.6	0.6	0.6	0.2	0	0
	3	0.2	0.6	0.8	0.6	0.2	0	0
	4	0.2	0.6	1.0	0.6	0.2	0	0

(c) OSC to DD

		\ DD						
		-3	-2	-1	0	1	2	3
	0	0	0	0.2	0.6	0.8	0.6	0.6
	1	0	0	0.2	0.6	1.0	0.8	0.6
OSC	2	0	0	0.2	0.6	0.8	1.0	0.8
	3	0	0	0.2	0.6	0.6	0.8	1.0
	4	0	0	0.2	0.6	0.6	0.6	1.0

(d) RSP to DP

		\ DP						
		-3	-2	-1	0	1	2	3
	0	0.2	0.4	0.4	0.6	0.6	0.8	0.6
	1	0.4	0.6	0.6	0.6	0.8	1.0	0.6
RSP	2	0.6	0.6	0.8	0.2	1.0	0.8	0.6
	3	0.6	0.8	1.0	0.6	0.8	0.6	0.6
	4	0.6	1.0	0.8	0.6	0.6	0.6	0.4

(e) RSP to DD

		\ DD						
		-3	-2	-1	0	1	2	3
	0	0	0	0.2	0.6	0.8	0.6	0.2
	1	0	0	0.2	0.6	1.0	0.6	0.2
RSP	2	0	0	0.2	0.6	1.0	0.6	0.2
	3	0	0	0.2	0.6	1.0	0.6	0.2
	4	0	0	0.2	0.6	0.8	0.6	0.2

FUZZY LOGIC CONTROL 89

(f) DIV to DP

		\multicolumn{7}{c}{DP}
		−3	−2	−1	0	1	2	3
DIV	0	0.2	0.6	0.8	0.6	0.2	0	0
	1	0.4	0.6	1.0	0.6	0.2	0	0
	2	0.6	0.6	1.0	0.6	0.2	0	0
	3	0.6	0.8	1.0	0.6	0.2	0	0
	4	0.6	1.0	0.8	0.6	0.2	0	0

(g) DIV to DI

		\multicolumn{7}{c}{DI}
		−3	−2	−1	0	1	2	3
DIV	0	0.2	0.4	0.4	0.4	0.2	0	0
	1	0.4	0.6	0.6	0.6	0.2	0	0
	2	0.6	0.6	0.8	0.6	0.2	0	0
	3	0.6	0.8	1.0	0.6	0.2	0	0
	4	0.6	1.0	0.8	0.6	0.2	0	0

(h) DIV to DD

		\multicolumn{7}{c}{DD}
		−3	−2	−1	0	1	2	3
DIV	0	0	0	0.2	0.6	0.8	0.6	0.6
	1	0	0	0.2	0.6	1.0	0.8	0.6
	2	0	0	0.2	0.6	0.8	1.0	0.8
	3	0	0	0.2	0.6	0.6	0.8	1.0
	4	0	0	0.2	0.4	0.6	0.6	1.0

(i) OFS to DP

		\multicolumn{7}{c}{DP}
		−3	−2	−1	0	1	2	3
OFS	0	0	0	0.2	0.6	0.8	0.6	0.2
	1	0	0	0.2	0.6	1.0	0.6	0.2
	2	0	0	0.2	0.6	0.8	0.6	0.6
	3	0	0	0.2	0.6	1.0	0.8	0.6
	4	0	0	0.2	0.6	0.8	1.0	0.6

(j) OFS to DI

		\multicolumn{7}{c}{DI}
		−3	−2	−1	0	1	2	3
OFS	0	0	0	0.2	0.6	0.8	0.6	0.6
	1	0	0	0.2	0.6	1.0	0.8	0.6
	2	0	0	0.2	0.6	0.8	1.0	0.8
	3	0	0	0.2	0.6	0.6	0.8	1.0
	4	0	0	0.2	0.4	0.6	0.6	1.0

condition with each column of a relation table (Table 3), take the lower value in each pair of compared elements, and then take the largest of the vector elements thus obtained. This result now must be defuzzified in order to obtain a crisp value for the tuning action. The centroid method [Equation (13)] is used here. Specifically, we weight the elements in the universe (strictly, the support set) of the action variable using the membership grades of the action, and then take the average by dividing the sum of these membership grades. The fuzzy decision table obtained by following these steps for every rule in the fuzzy rule base is given in Table 4. It has been decided to take no action when the conditions are satisfactory, even though an optimal tuning strategy would suggest some other action under these conditions. The relation used for updating a PID parameter is

$$p_{new} = p_{old} + \Delta p (p_{max} - p_{min}) / p_{sen} \qquad (18)$$

in which p denotes a PID parameter. The subscript "*new*" denotes the updated value and "*old*" denotes the previous value. The incremental action taken by the fuzzy controller is denoted by Δp. Upper and lower bounds for a parameter are denoted by the subscripts *max* and *min*. A sensitivity parameter p_{sen} is also introduced for adjusting the sensitivity of tuning, when needed. A fuzzy tuner of this type will be used in the application, which is described in Chapter 6.

When using the decision table (Table 4) it is assumed that an intelligent preprocessor is available to monitor the process behavior and determine the condition of the process attributes (OSC, RSP, DIV, and OFS) as fuzzy states rather than numerical values. Alternatively, it is also possible to construct a decision table in which the conditions are expressed as discrete numerical values. Then, either some form of interpolation or rounding-off to the nearest discrete value would be necessary.

■ SUMMARY

In this chapter the basics of fuzzy logic control are presented. The main steps of fuzzy control are the off-line steps: development of a rule base, establishment of the membership functions for various fuzzy states in the condition and action variables, and the development of a multidimensional membership function for the rule base (using *min, max* operations). The on-line steps are *fuzzification* of the process measurements, application of the *compositional rule of inference* (*sup-min*) to determine fuzzy inference, and *defuzzification* of the inference to obtain the crisp control action. The entire procedure may be simplified through the use of fuzzy singletons for process

TABLE 4
FUZZY DECISION TABLE FOR A SERVOCONTROLLER

		Action	
Condition	DP	DI	DD
NON	0.0	0.0	0.0
OSC = LOW	−1.412		
= MOD			
= HIG			
= VHI			
RSP = LOW			
= MOD			
= HIG			
= VHI			
DIV = LOW			
= MOD			
= HIG			
= VHI			
OFS = LOW			
= MOD			
= HIG			
= VHI			

Exercise: Using the procedure above, fill in the remaining entires of the decision table.

measurements. Thus, it is not necessary to first determine a multidimensional membership function of the rule base. Instead, each rule is considered separately (for the *min* operation with singletons), and the resulting partial (rule-by-rule) inferences are superimposed (the *sup* or *max* operation) to determine the overall control inference. The result is defuzzified to determine the crisp control action.

Several methods are available for defuzzification. The *centroid method* uses the centroid of the membership function of the control inference. The *mean of maxima method* first detects the peak values of the membership function and then computes their mean (typically weighted by the corresponding membership grades). Both methods use the membership function of control inference in the entire support set. The two methods may be modified to use a *thresholded support set* (an α-cut) in which part of the boundary elements is left out and only a core section of the membership function is used. The centroid method is by far the most commonly used and is known to be robust, less sensitive to changes, and produces a more stable system response. The method can be more sensitized by thresholding. The mean of maxima method is relatively unaffected by thresholding.

Two architectures are commonly used in fuzzy logic control. More common is the structure in which the fuzzy controller is within the *low-level control loop* and it performs direct control functions. This architecture has some drawbacks with respect to speed, accuracy, and sensitivity, and as a result, at least some of the following conditions must be met if this approach is to be effective. Process time constants and required control bandwidth are high; plant is complex and ill-defined, without the feasibility of analytical or experimental modeling; and experience in low-level, human-in-the-loop control is available and may be expressed as fuzzy, linguistic protocols. The second architecutre is *hierarchical*, and the fuzzy decision making system occupies a higher layer above the direct control level. Here, a conventional crisp controller is used for direct control. A *signal preprocessor* is needed because high-level decision making typically needs high-level information. The performance of the system is monitored using the preprocessed sensory information. Typical high-level control functions performed on this basis include tuning, adaptation, self-organizing, and dynamic restructuring of the control system.

■ PROBLEMS

1. Two fuzzy sets A and B are represented by the membership functions shown below:

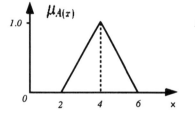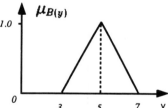

   Sketch the membership functions of the following fuzzy sets/relations:
   (a) $A \cup B$.
   (b) $A \cap B$.
   (c) NOT A.
   (d) $A \rightarrow B$.

   In (a) and (b) assume that A and B are defined in the same universe $X = [0, 8]$, and in (d) assume that B is defined in another universe Y. Both universes are subsets of the real line $\Re$.

   In case (d) suppose that a crisp reading of A' is observed as $x = 5$. What is the membership function of the corresponding inference B'? If this inference is defuzzified using the centroid method, what is the crisp action?

2. Consider the experimental setup of an inverted pendulum, as described in de Silva, 1989 and shown in Figure 7.

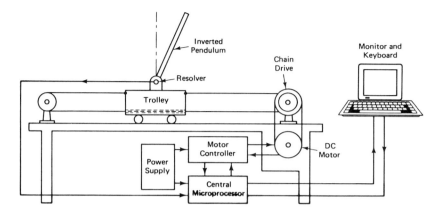

Figure 7 A computer-controlled inverted pendulum.

Instead of statistical process control, suppose that direct fuzzy logic control is used to keep the inverted pendulum upright. The process measurements are the angular position, about the vertical (*ANG*) and the angular velocity (*VEL*) of the pendulum. The control action (*CNT*) is the current of the motor driving the positioning trolley. The variable *ANG* takes two fuzzy states: positive large (*PL*) and negative large (*NL*). Their memberships are defined in the support set [-30°, 30°] and are trapezoidal. Specifically,

$$\mu_{PL} = 0 \quad \text{for} \quad ANG = [-30°, -10°]$$
$$= \text{linear } [0, 1.0] \quad \text{for} \quad ANG = [-10°, 20°]$$
$$= 1.0 \quad \text{for} \quad ANG = [20°, 30°]$$
$$\mu_{NL} = 1.0 \quad \text{for} \quad ANG = [-30°, -20°]$$
$$= \text{linear } [1.0, 0] \quad \text{for} \quad ANG = [-20°, 10°]$$
$$= 0 \quad \text{for} \quad ANG = [10°, 30°]$$

The variable *VEL* takes two fuzzy states *PL* and *NL*, which are defined quite similarly in the support set [–60°/s, 60°/s]. The control inference *CNT* can take three fuzzy states: positive large (*PL*), no change (*NC*), and negative large (*NL*). Their membership functions are defined in the support set [–3A, 3A] and are either trapezoidal or triangular. Specifically,

$$\mu_{PL} = 0 \quad \text{for} \quad CNT = [-3A, 0]$$
$$= \text{linear } [0, 1.0] \quad \text{for} \quad CNT = [0, 2A]$$
$$= 1.0 \quad \text{for} \quad CNT = [2A, 3A]$$
$$\mu_{NC} = 0 \quad \text{for} \quad CNT = [-3A, -2A]$$
$$= \text{linear } [0, 1.0] \quad \text{for} \quad CNT = [-2A, 0]$$
$$= \text{linear } [1.0, 0] \quad \text{for} \quad CNT = [0, 2A]$$
$$= 0 \quad \text{for} \quad CNT = [2A, 3A]$$
$$\mu_{NL} = 1.0 \quad \text{for} \quad CNT = [-3A, -2A]$$
$$= \text{linear } [1.0, 0] \quad \text{for} \quad CNT = [-2A, 0]$$
$$= 0 \quad \text{for} \quad CNT = [0, 3A]$$

Four fuzzy logic rules are used in control:

```
        If    ANG  is  PL   and  VEL  is  PL   then  CNT  is  NL
else    if    ANG  is  PL   and  VEL  is  NL   then  CNT  is  NC
else    if    ANG  is  NL   and  VEL  is  PL   then  CNT  is  NC
else    if    ANG  is  NL   and  VEL  is  NL   then  CNT  is  PL
end     if.
```

(a) Sketch the four rules in a membership diagram for the purpose of making control inferences.
(b) If process measurements of $ANG = 5°$ and $VEL = 15°/s$ are made, indicate on your sketch the corresponding control inference.

3. Consider again the comfort control system shown in Figure 2. Suppose that some of the rules in the control knowledge base are

```
        If    T  is  ZR   and  H  is  ZR   then  C  is  NC
else    if    T  is  PL   and  H  is  ZR   then  C  is  PL
else    if    T  is  ZR   and  H  is  PL   then  C  is  PL
else    if    T  is  PL   and  H  is  PL   then  C  is  PL
end     if …
```

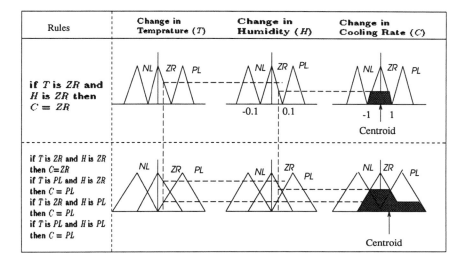

Figure 8 The effect of the overlap in membership functions on the accuracy of control inference.

Note that T = change in temperature from reference, H = change in humidity from reference, and C = change in cooling rate from reference. Two cases can be considered:
(a) The membership functions do not overlap.
(b) The membership function have some overlap.

These two cases are illustrated in Figure 8. Discuss why case (b) provides more realistic control inferences than case (a). You may consider the inference C produced in the two cases for a specific reading of T and H.

4. Many recent papers on fuzzy logic control use simple, linear models to represent the plants, thereby contravening the purpose of using knowledge-based control in the first place (de Silva, 1993a). In one such paper the context variables of the control rule base are taken as the response error, the summation of the error, and the change in error, somewhat analogously to the classic PID control. The work goes on to assume that the behavior of the control system would be similar to what one normally observes under PID control, for example, in terms of stability, speed of response, and steady-state accuracy. Discuss the validity of such conclusions.

5. Considering an intelligent controller as a real-time expert system, briefly explain the following terms:
   (a) Knowledge Engineering Interface
   (b) Knowledge Validation
   (c) Explanation of Reasoning
   (d) Truth Maintenance
   (e) Dynamic Knowledge Base
   (f) Real-Time Performance
   (g) Uncertainty Management

6. Knowledge based techniques are useful in injection molding of material. In particular, the following three situations may be considered:
   (i) Expert system to assist the development of molds.
   (ii) Intelligent controller for the injection molding process.
   (iii) Knowledge-based supervisor to monitor the process performance (e.g., deflects in the molded parts) and to take corrective actions.

   In (i) mold geometry, gate location, vent locations, etc., may be considered. In (ii) material characteristics, mixing, mold orientation, pressure, temperature, etc., are important factors. In (iii), problems such as poor density, hardness, and strength, and also nonuniformities and deformations in the molded parts are of importance. Discuss distinctive characteristics of the above three categories of application; and indicate whether fuzzy logic could be effectively used in each case.

7. Does "fuzzification" of a crisp item of data generally lead to loss of information? Carefully justify your answer. You may use a statistical analogy.

8. High-speed, digital, fuzzy processor chips are commercially available with inference processing speeds of 650 μs per cycle (with 20 rules). The function list of one such processor provides information on the following features:
   1. Inference method
   2. Defuzzification method
   3. Number of I/O ports
   4. Rule format
   5. Number of rules
   6. Rule weighting
   7. Antecedent membership functions
   8. Consequent membership functions

   Give the meaning of each of these terms.

FUZZY LOGIC CONTROL 97

9. Consider the freeway incident-detection problem as outlined in [Hall, L., Hall, F.L., and Shi, Y (1991). A Catastrophe Theory Approach to Freeway Incident Detection, *Proc. 2nd Int. Conf. on Applications of Advanced Technologies in Transportation Engineering*, New York, NY, pp. 373–377]. Use the following definitions:

Flow rate f = number of vehicles passing a specified point of a lane, per unit time

Occupancy $d = \dfrac{\text{Number of vehicles in a given segment of a lane} \times 100\%}{\text{Maximum number of vehicles the segment can occupy}}$

Note that discrepancies can result if the vehicle parameters (geometric and kinematic) and road parameters (road conditions, grade, etc.) are not uniform or fixed.

It is required to establish a lower-bound curve for uncongested flow on the Occupancy-Flow Rate plane such that, points above the curve represent uncongested flow and the points below represent congested flow. The figure shows a schematic representation of a set of experimental data (occupancy, flow) pairs for a lane segment of freeway. Here u denotes a data point of uncongested flow and c denotes one of congested flow. It is noted that some overlap is present, and hence it is difficult to establish a single lower-bound curve.

(a) Give reasons for such overlaps.

Suppose that a fuzzy representation of the lower-bound curve is considered within the overlap band. Specifically, curves of the form

$$f = bd^{1/n} + a \tag{1}$$

are considered with fixed parameters b and n (say, $b = 6.7$ and $n = 0.5$) and the parameter a being fixed for each curve. The units of f are vehicles/min, and d is expressed as a percentage. It is required to estimate a membership function $\mu_U(a)$ such that the membership grade for a particular value of a would represent the possibility that the corresponding curve given by Equation (1) falls into the uncongested region. In particular, if $\mu_U(a) = 0$ the curve would be completely in the congested region, and if $\mu_U(a) = 1$ it would be completely in the uncongested region.

(b) Suggest a way to estimate this membership function, using the experimental data that are schematically shown in the figure, say for values of *a* in the range [–70, 20].

(c) Discuss whether your estimate can be more appropriately considered as either a *belief function* or a *plausibility function* rather than a *membership function*. (See Appendix A).

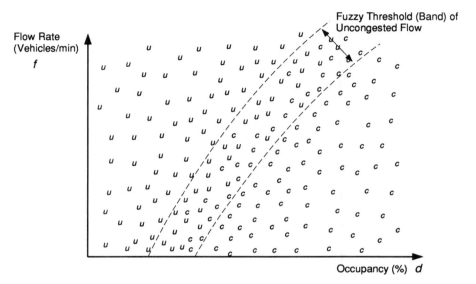

Figure 9 A schematic representation of occupancy-flow experimental information for a freeway lane segment

10. A fuzzy-logic rule, for example, as given below

$$R_i:(A_i, B_i) \rightarrow C_i$$

is called a "fuzzy association". A *fuzzy associative memory* (FAM) is formed by partitioning the universe of discourse of each condition variable (i.e., A_i and B_i in the above example) according to the level of *fuzzy resolution* chosen for these antecedents, thereby generating a grid of FAM elements. The entry at each grid element in the FAM corresponds to the fuzzy action (C_i in the above example). A fuzzy associate memory may be, then, interpreted as a geometric or tabular representation of a fuzzy-logic rule base.

A fuzzy associate memory of a fuzzy-logic controller is shown in the FAM diagram in Table 5. Express these rules in a linguistic form. What do the empty entries in the table (matrix) represent?

FUZZY LOGIC CONTROL 99

11. In an attempt to compare the "centroid method" and the "mean of max method" of defuzzification, suppose that the control inference C' has a triangular membership function with $\mu_{C'}(c)$ with a support set $[-c_o, c_o]$ and a peak value of 1.0 at $c = c_o/2$. A thresholded defuzzification is used with a threshold value of α. Show that the defuzzified control action $\hat{c}$ is given by

$$\hat{c} = \frac{c_0}{2} \text{ for all } \alpha, \text{ in the mean of max method}$$

$$= \frac{c_0}{6}(1 + 2\alpha) \text{ in the centroid method}$$

TABLE 5
FAM DIAGRAM OF A FUZZY-CONTROL RULE BASE

		\multicolumn{5}{c}{Change in Error Δe}
		NL	NS	ZO	PS	PL
	PL		NS	NL	NL	NL
Error	PS		ZO	NS	NL	NL
e	ZO	PL	PS	ZO	NS	NL
	NS	PL	PL	PS	ZO	
	NL	PL	PL	PL	PS	

Note: Here, N denotes "negative"; P denotes "positive", S denotes "small"; L denotes "large", and ZO denotes "zero".

12. This problem deals with rule interaction (see Appendix A) in fuzzy logic control. First consider just one rule,

$$R_1 : A_1 \to C_1$$

expressed in the discrete form:

$A_1 = 0.7/a_1 + 0.6/a_2 + 0.1/a_3$ with cardinality 3

$C_1 = 0.5/c_1 + 0.4/c_2$ with cardinality 2

Show that this rule base may be expressed as

$$\mu_{R_1}(a_i, c_j) = \begin{bmatrix} 0.5 & 0.4 \\ 0.5 & 0.4 \\ 0.1 & 0.1 \end{bmatrix}$$

Now supose that a fuzzy observation

$$A_1' = 0.7/a_1 + 0.6/a_2 + 0.1/a_3$$

is given. Show by applying the *max-min composition* that the corresponding control inference is given by

$$C' = 0.5/c_1 + 0.4/c_2$$

Next suppose that a second rule R_2 is available,

$$R_2: A_2 \rightarrow C_2$$

with,

$$A_2 = 0.6/a_1 + 0.7/a_2 + 0.2/a_3$$
$$C_2 = 0.4/c_1 + 0.5/c_2$$

Show that the membership function of the combined rule base $R = R_1 V R_2$ is given by

$$\mu_R(a_i, c_j) = \begin{bmatrix} 0.5 & 0.5 \\ 0.5 & 0.5 \\ 0.2 & 0.2 \end{bmatrix}$$

Now, for a fuzzy observation

$$A_1' = 0.7/a_1 + 0.6/a_2 + 0.1/a_3$$

which is the same as before, show that the corresponding control inference is

$$C' = 0.5/c_1 + 0.5/c_2$$

which is different from the previous inference.

REFERENCES

de Silva, C.W. (1989). *Control Sensors and Actuators*, Prentice-Hall, Englewood Cliffs, NJ.

de Silva, C.W. (1993a). Soft Automation of Industrial Processes, *Eng. Appl. Artif. Intelligence*, Vol. 6, No. 2, pp. 87–90.

de Silva, C.W. (1993b). Knowledge-Based Dynamic Structuring of Process Control Systems, *Proc. 5th Int. Fuzzy Systems Assoc. World Congress*, Seoul, Korea, Vol. II, pp. 1137–1140.

de Silva, C.W. and MacFarlane, A.G.J. (1988). Knowledge-Based Control Structure for Robotic Manipulators, *Proc. IFAC Workshop on Artificial Intelligence in Real-Time Control*, Swansea, Wales, pp. 143–148.

de Silva, C.W. and MacFarlane, A.G.J. (1989a). *Knowledge-Based Control with Application to Robots*, Springer-Verlag, Berlin.

de Silva, C.W. and MacFarlane, A.G.J. (1989b). Knowledge-Based Control Approach for Robotic Manipulators, *Int. J. Control*, Vol. 50, No. 1, pp. 249–273.

Dubois, D. and Prade, H. (1980). *Fuzzy Sets and Systems*, Academic Press, Orlando, FL.

Mamdani, E.H. (1977). Application of Fuzzy Logic to Approximate Reasoning Using Linguistic Synthesis, *IEEE Trans. Comp.*, Vol. C-26(12), pp. 1182–1192.

Pedrycz, W. (1989). *Fuzzy Control and Fuzzy Systems*, Research Studies Press, Ltd., Somerset, England.

Procyk, T.J. and Mamdani, E.H. (1979). A Linguistic Self-Organizing Controller, *Automatica*, Vol. 15, pp. 15–30.

Sugeno, M. (Ed.) (1985). *Industrial Applications of Fuzzy Control*, North-Holland, Amsterdam.

Zadeh, L.A. (1975). The Concept of a Linguistic Variable and Its Application to Approximate Reasoning, Parts 1-3, *Inf. Sci.*, pp. 43–80, 199–249, 301–357.

5 KNOWLEDGE-BASED TUNING

■ INTRODUCTION

In knowledge-based control the control signals are generated by an appropriate inference mechanism, and employing a control knowledge base which is typically expressed as a set of rules. Unlike hard-algorithmic control, knowledge-based control does not primarily depend on analytic control algorithms or on accurate models of the plant. Clearly, knowledge-based control is particularly attractive when considerable knowledge, expertise (or specialized knowledge), and experience are available in controlling a particular plant, and when the plant is rather incompletely known or complex to be accurately modeled. Fuzzy control, which is based on the fuzzy logic of Zadeh (1965), is a class of knowledge-based control that incorporates control knowledge in the form of a set of linguistic rules. These rules may contain *fuzzy*, *noncrisp* or *soft* terms such as "rather high", "slightly lower", "very slow", and "accurate". It should be noted that *vagueness*, *ambiguity*, *generality*, *imprecision*, and *uncertainty* are not exactly synonymous with fuzziness. Examples of the first five situations are given by the following statements: "I will read the paper some day", "I may or may not read the paper", "I will read x papers in y months", "I will read the paper within the next twenty-four hours", and "There is a 50 percent probability that I will read the paper within the next twenty-four hours". Instead, the statement, "I will read the paper soon" is fuzzy, as it contains the fuzzy term "soon". There are numerous practical implementations of fuzzy logic control. In Japan, in particular, consumer products and utilities such as washing machines, vacuum cleaners, toasters, hand-jitter-compensated video cameras, television sets which can automatically adjust brightness and volume depending on room conditions, and subway trains that use this method of control are already commercially available.

Conventional fuzzy control uses linguistic rules of the typical form, "If the error is positive and large and the error rate is negative and small, then slightly increase the control signal". To come up with such rules the control engineer must have either a thorough insight of the dynamics of the overall control system, including the plant, or tremendous experience in actually applying various types of control signals to the system and observing the system response. Because knowledge-based control is intended for a complex or poorly known plant it is not realistic to assume that the dynamic behavior of the system is completely known. For a moderate- or high-bandwidth process such as robot (here, the term bandwidth refers to the speed or maximum frequency of response; de Silva, 1989), it is also unrealistic to manually sequence the control signals and observe the system behavior on-line in order to acquire the necessary knowledge and experience for establishing a set of fuzzy control rules. On the other hand, it is much easier for a human expert to manually tune a control system of moderate-to-high bandwidth during operation, and thereby gather "tuning knowledge". Manual tuning is still widely used in process control practice.

Ziegler-Nichols-type empirical and heuristic tuning rules (Ziegler and Nichols, 1942) are known to be quite effective for a variety of servo systems, and as a result many commercial systems of servo tuning based on the Ziegler-Nichols method are still available. Much of the practical experience in tuning a servo controller is available not as hard (crisp) algorithms based on accurate models and rigorous analysis, but rather in the form of "soft" rules. These reasons and facts led us to consider a particular type of control structure (de Silva and MacFarlane, 1989) in which fuzzy logic is used not directly in control, but rather at a higher level, for tuning the direct controllers that are located at a lower level. In this structure crisp control algorithms or high-bandwidth servo controllers (bandwidth refers here to the frequency of control cycles) serve as a direct controller, and a knowledge-based tuner of relatively low bandwidth occupies a higher level to monitor and tune the direct controllers. This structure naturally conforms to the fact that experience-based tuning protocols are inherently fuzzy in general. It has another main advantage in that fuzzy schemes are relegated to a higher level so that they do not enter into the direct control loops, thereby maintaining the control bandwidth (or speed of control) high and perhaps improving the accuracy of control as well. The specific control structure treated by us may be compared to the general structure of "expert control" as introduced by Astrom et al. (1986). In the latter control structure the system has several algorithms for the purposes of control, supervision and adaption. An expert system decides when a particular algorithm should be used. It should be emphasized that our purpose in the present context is to automate the human expertise of the control engineer who is not the plant operator except in special circumstances.

The objective of this chapter is to build an analytical framework for a knowledge-based tuner of the type mentioned above. In particular, the problem of tuning a servo controller in which a set of linguistic tuning protocols is available as the knowledge base is considered. First, fuzzy logic is used to represent the tuning knowledge base in the form of a membership function of a fuzzy relation, and this is extended to the tuning problem through the application of the compositional rule of inference. Next, the concepts of *rule dissociation*, *fuzzy resolution*, and *resolution relationships* are introduced to build an analytical framework for the problem. Stability and computational considerations of the tuning system are investigated. Finally, an example of tuning a proportional-integral-derivative (PID) servo for a nonminimum-phase process with transport delay is presented to illustrate the application of the concepts discussed in this chapter.

■ THEORETICAL BACKGROUND

There are many methods of controller tuning and adaptive control. A good description of available techniques is found in Astrom and Wittenmark (1989). For example, a model-referenced adaptive system consists of an inner loop of direct feedback control and an outer loop of parameter adjustment, and the parameters are adjusted such that the response of the plant is forced toward the response of the reference model. In an original adaptation algorithm known as the *MIT rule*, the parameters are changed in the direction opposite to the gradient vector of the quadratic error. In this chapter the desired performance is expressed not in terms of a reference model, but rather in terms of a set of performance attributes that can be directly correlated to the human expertise of a control engineer. Furthermore, the tuner, which is placed at an upper level in the control structure, uses a set of linguistic fuzzy rules (a "soft" algorithm) instead of the "hard" algorithms used in conventional adaptive control.

This section establishes an analytical foundation for the fuzzy tuning problem. A parallel is drawn between the knowledge-based tuning problem and the parameter-adaptive control problem. In the general problem of fuzzy tuning the inferences are made using the compositional rule of inference (Zadeh, 1979). This has been fully addressed in Chapters 3 and 4. The general problem is simplified by first assuming the case of system-characteristic adaptation with respect

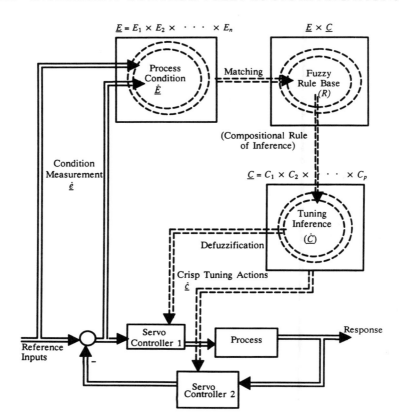

Figure 1 The structure of a knowledge-based servo tuner.

to a test input (Narendra and Annaswamy, 1989), and then using a preliminary dissociation of the tuning rule base (de Silva and MacFarlane, 1989).

Consider the general structure of a fuzzy tuner as shown in Figure 1. The control system is represented by the state-space error model:

$$\dot{\mathbf{e}}(t) = \mathbf{f}(\mathbf{e}(t), \mathbf{c}(t), \mathbf{u}(t)) \tag{1}$$

where:
  time $t \in \Re^+$
  the error state vector $\mathbf{e}(t) : \Re^+ \to \Re^n$
  the tuning parameter vector $\mathbf{c}(t) : \Re^+ \to \Re^p$
  the reference input vector $\mathbf{u}(t) : \Re^+ \to \Re^d$
  the mapping $\mathbf{f} : \Re^n \times \Re^p \times \Re^d \to \Re^n$

Two servo-control modules, one in the forward path of the control system and the other in the feedback path, are present.

Rule Base

Generally, both the input vector **u** and the error vector **e** must be measured (the measurements being denoted by **u**′ and **e**′) in order to determine a fuzzy tuning inference for the parameter vector **c** (the tuning actions being denoted by **ĉ**), as for a general parameter-adaptive control algorithm. The rule base of tuning for this general case will contain conditions for both **e** and **u** and corresponding actions for **c**. We restrict our analysis to the case of tuning in which the desired performance is specified in terms of the error response to a step input. The test input may be applied during a quiescent period of the system and the error response monitored. By adopting this procedure, it is possible to simplify the tuning rule base to include conditions (antecedents) of **e** only. Furthermore, it is not necessary to measure **u** in this case. While it has these advantages, the use of a step input for establishing the "context" of tuning is clearly a limitation. In this case the tuning must be restricted to quiescent periods of the plant. However, the concepts presented here may be extended to other types of tuning inputs and for tuning under nonquiescent plant conditions, provided a corresponding knowledge base of tuning is available. Also, the observed attributes may not necessarily be the error state **e**′ itself, but some vector **ê** that is related to **e**′. This approach is quite akin to manual tuning, e.g., via the Ziegler-Nicholas approach, and therefore, further justified in view of the fact that the rule base that is used in fuzzy-logic tuning is assumed to originate and evolve through human experience. In contrast, parameter identification based on input-output data is done not through human experience of tuning, but rather by using sophisticated analytical relations and algorithms that cannot be realistically extended in general to a linguistic knowledge base for use in fuzzy tuning. Nevertheless, such an extension should not be completely ruled out.

Consider a tuning rule base that is given by the general form

$$\underset{i,j,\cdots,k}{\text{ELSE}}[\text{IF } E_1^i \text{ THEN IF } E_2^j \text{ THEN IF } \cdots E_n^k \text{ THEN } C_1^a \text{ AND } \cdots C_p^c] \quad (2)$$

Here, E and C denote the "fuzzy" variables of error condition and tuning action, respectively. The attached subscript is the index that identifies a particular variable in a vector of variables. A fuzzy variable can assume several states (e.g., very small, small, medium, large, and very large). The superscript attached to a fuzzy variable is the index that identifies a particular state of the variable. The lower case letters e and c denote "crisp" quantities (e.g., actual measurements or actual tuning values) associated with E and C. Suppose that there are n error-attribute variables $e_1, e_2, \ldots, e_n$ and p tuning-action variables $c_1, c_2, \ldots, c_p$. Note that in general each error attribute (antecedent) e_s may assume m discrete fuzzy states $E_s^1, E_s^2, \cdots, E_s^m$ for

$s = 1, 2, \ldots, n$, and similarly, each tuning action (consequent) c_q may assume r discrete fuzzy states $C_q^1, C_q^2, \ldots, C_q^r$ for $q = 1, 2, \ldots, p$. This concept of *"fuzzy resolution"*, which is discussed in more detail later, is introduced here. Specifically, the integer value m represents the fuzzy resolution of E_s, and similarly r represents the fuzzy resolution of C_q. The tuning rules themselves are assumed to be available in a linguistic form, a typical tuning rule being "If the response is moderately oscillatory then slightly decrease the proportional gain and slightly increase the derivative time constant". The fuzzy variables E_s and C_q are completely represented by their membership functions (Zadeh, 1965; Dubois and Prade, 1980)

$$\mu_{E_s}(e_s) : \Re \to [0,1] \quad \mu_{C_q}(c_q) : \Re \to [0,1]$$

The rule base contains fuzzy-logical operations such as *implication*, *complement*, *union*, and *intersection*. Then, by employing suitable norms for these operations it is possible to represent the rule base (Equation (2)) by a fuzzy relation R whose membership function is $\mu_R(\mathbf{e}, \mathbf{c})$. The four logical operations that are useful to us and the most commonly used norms (i.e., mathematical functions) for these operations are listed below:

Complement (Not):	$\mu_{\bar{A}}(x) \triangleq 1 - \mu_A(x)$	$\forall x \in X$
Disjunction (Or):	$\mu_{A \vee B}(x) \triangleq sup(\mu_A(x), \mu_B(x))$	$\forall x \in X$
Conjunction (And):	$\mu_{A \wedge B}(x) \triangleq min(\mu_A(x), \mu_B(x))$	$\forall x \in X$
Implication (If-Then):	$\mu_{A \to B}(x, y) \triangleq min(\mu_A(x), \mu_B(y))$	$\forall x \in X, \forall x \in X$

Here, X is the universe of discourse in which the various fuzzy variables (A, $\bar{A}$, and B) are defined, except in the case of implication, in which the consequent B is defined in a separate universe of discourse Y. These definitions result in the so-called sup-min composition, as discussed in Chapters 3 and 4. Other norms may be used, leading to other types of composition with their characteristic advantages and disadvantages. For example, for fuzzy intersection, one may use the *product*, $[\mu_A(x)\mu_B \cdot (x)]$ or the *bounded max* (bold intersection) $max[0, \mu_A(x) + \mu_B(x) - 1]$. These two and the previously stated min norm fall into a class known as *triangular norm* (or *T-norm*) which is a function $T: [0,1] \times [0,1] \to [0,1]$ that satisfies certain conditions (Dubois and Prade, 1980). More details are found in Appendix A. Alternative norms for fuzzy union include the set addition, $[\mu_A(x) + \mu_B(x) - \mu_A(x) \cdot \mu_B(x)]$ and the *bounded min* (bold union), $min[1, \mu_A(x) + \mu_B(x)]$. A bounded-min norm that is sometimes used for implication is $min[1, 1 - \mu_A(x) + \mu_B(y)] \, \forall \, x \in X, \, \forall \, y \in Y$, which is akin to crisp logical implication of the conventional type. Even though this definition of fuzzy implication carries a strong intuitive appeal, the previous one is much simpler and often provides more accurate results. Note that the actual numerical value for a tuning action can depend on the particular mathematical norms that are employed for various operations in the rule base. Past experience and judgment are useful in selecting a norm for an operation. The formulation presented in this chapter does not depend on the choice of the norms, and for illustrative purposes we have employed the most commonly used norms.

To use the sup-min composition in the overall rule base [Equation (2)] we first define a useful notation. The set of error attributes $\{E_1, E_2, \ldots, E_m\}$ is denoted by $\mathbf{E}$. These attributes may respectively assume the fuzzy states $i, j, \ldots, k$, and these states are collectively denoted by the state $\mathbf{i}$. The fuzzy set of E_s has a support set (i.e., the subset corresponding to nonzero values of the membership function), and the variable over this support set is e_s. Then, the set of membership functions of $\mathbf{E}$ which assume the fuzzy state $\mathbf{i}$ is given by

$$\mu_{\mathbf{E}^i}(e) \triangleq [\mu_{E_1^i}(e_1), \mu_{E_2^j}(e_2), \cdots, \mu_{E_n^k}(e_n)] \qquad (3)$$

For each antecedent (condition) state $\mathbf{i}$ of the set of error attributes $\mathbf{E}$, according to the rule base [Equation (2)], there is a corresponding consequent (action) state $\mathbf{j}$ for the tuning parameters $\mathbf{C} \triangleq \{C_1, C_2, \ldots, C_p\}$. Using the same definition as in Equation (3), the associated set of membership functions is denoted by $\mu_{\mathbf{C}^j}(\mathbf{c})$.

Note that for a particular condition state $\mathbf{i}$ of the error attributes there will be a unique action state $\mathbf{j}$, assuming, of course, that there are no conflicting rules. However, some of the indices in the index set $\mathbf{j}$ may represent a "no change" state because not all the tuning parameters are involved in a given rule. The associated membership values are taken to be unity in this generalized notation. Extending this further, all attributes may not be associated in a particular rule. Hence, the set $\mathbf{i}$ may include error attributes that do not exist in a particular rule, and as before, the corresponding membership values are taken to be unity. With this notation, the membership function of the rule base [Equation (2)] is given by

$$\mu_R(\mathbf{e}, \mathbf{c}) = \sup_{i,j} \min[\mu_{\mathbf{E}^i}(\mathbf{e}), \mu_{\mathbf{C}^j}(\mathbf{c})] \qquad (4)$$

If E and C are used not only as labels of fuzzy variables, but also to denote the support sets of the corresponding variables, it should be clear that the rule base μ_R is defined in the Cartesian product space $\mathbf{E} \times \mathbf{C} \triangleq E_1 \times \cdots \times E_n \times C_1 \times C_2 \times \cdots \times C_p$, as indicated in Figure 1.

The first stage of rule dissociation is introduced now. Specifically, if a rule proposes actions on more than one tuning parameter, it is quite valid to consider the action parameters sequentially. This does not imply that the coupling (or interaction) among the condition and the action variables is neglected, but rather it is considered in a sequential manner. The problem of rule base decoupling is addressed separately, in Chapter 7. The dissociation of the rules has a secondary benefit as well. Specifically, if the actions can be prioritized according to the order of importance, it will be desirable to arrange the tuning sequence in the same order. For example, in a PID controller it is appropriate to arrange the tuning actions in the order D-P-I which is the descending order of the *natural speed of reaction* of these three control actions. In this manner the rule base (Equation (2)) can be partitioned in such a way that each partition is uniquely associated with one tuning parameter. Then there is a one-to-one correspondence between the rule base partitions and the tuning parameters. For instance, the rule base partition that corresponds to the qth tuning parameter has the membership function

$$\mu_{R_q}(\mathbf{e}, c_q) = \sup_i min[\mu_{\mathbf{E}^i}(\mathbf{e}), \mu_{C_q^b}(c_q)] \quad \text{for } q = 1, 2, \cdots, p \quad (5)$$

which should be compared to Equation (4). Note that the index b which denotes the fuzzy state of C_q in Equation (5), as for $\mathbf{j}$ in Equation (4), is completely determined by the state $\mathbf{i}$ of the error attributes, and as a result, b does not enter as an independent index.

Compositional Rule of Inference

As shown in Figure 1, suppose that a set of attribute observations are made related to the system behavior. These are crisp quantities in general and, for the purpose of the subsequent fuzzy analysis, they must be fuzzified (Mamdani, 1977; Dubois and Prade, 1980), as discussed in Chapter 4. If crisp quantities are represented by "fuzzy singletons" with unity membership grade at the crisp value and zero grade elsewhere in the support set, no information will be lost in fuzzification of a crisp quantity. The resulting fuzzy observations $\hat{\mathbf{E}} \triangleq \{\hat{E}_1, \hat{E}_2, \ldots, \hat{E}_n\}$ must be matched with the rule base R in order to arrive at a fuzzy tuning inference $\hat{\mathbf{C}} \triangleq \{\hat{C}_1, \hat{C}_2, \ldots, \hat{C}_n\}$. This is accomplished by applying the compositional rule of inference (Zadeh, 1979), which was introduced in Chapters 3 and 4. Specifically, as in Equations (4) and (5), the *sup-min* composition is used. It may be easily verified, then, that the membership function of the fuzzy inference for the qth parameter is given by

$$\mu_{\hat{C}_q}(c_q) = \sup_{\mathbf{e}} min[\mu_{\hat{\mathbf{E}}}(\mathbf{e}), \mu_{R_q}(\mathbf{e}, c_q)] \quad \text{for } q = 1, 2, \cdots, p \quad (6)$$

Because the tuning actions themselves must be crisp, these fuzzy tuning inferences must be "defuzzified" using one of the available methods (Dubois and Prade, 1980). As stated in Chapter 4, the *center of gravity*, or *centroid method* is widely used, and is given by

$$\hat{c}_q = \frac{\int_C c\mu_{\hat{c}_q}(c)dc}{\int_C \mu_{\hat{c}_q}(c)dc}] \quad \text{for } q = 1,2,\cdots,p \tag{7}$$

in which the integration is carried out over the support set of C. Tuning the servo controllers (Figure 1) is done using the tuning actions $\hat{\mathbf{c}} \triangleq \{\hat{c}_1, \hat{c}_2, \ldots, \hat{c}_q\}$. Obviously, some information is lost through defuzzification, but this is an essential practical reality. Furthermore, the particular defuzzification procedure as given by Equation (7) generally tends to make the knowledge-based tuner more robust. While a mathematical proof cannot be given for this statement, the robustness of the center of gravity method has been experienced in practical applications of fuzzy control. Intuitively, because the integration associated with the computation of the center of gravity is an averaging operation, any localized errors or disturbance tend to filter out in the process and will contribute to the robustness of the tuner.

■ ANALYTICAL FRAMEWORK

The rule base [Equation (2)], in its state-space form, has the disadvantage that it is expressed in terms of the *error states* as the condition variables. This will inevitably make the rules coupled and rather complex. Furthermore, because knowledge-based tuning relies heavily on human experience and the expertise of the control engineer, it is not advantageous to limit the observed attributes to error state variables alone. Often, other attributes such as *percentage overshoot, steady-state error, delay time,* and *rise time* serve as convenient antecedents in the experience of manual tuning of a servo controller, and this is particularly true when the tuning attributes are observed in response to a test input. The use of such familiar performance attributes will not only make the knowledge base used in tuning more realistic and accurate, but also will uncouple (again, through human expertise) the rules in the knowledge base, which can significantly improve the computational efficiency. The concept of rule dissociation and the introduction of resolution relations to link the membership functions of the condition variables with those of the action variables form the basis of the analytical framework presented here.

Rule Dissociation

In manual tuning of a control system a human expert tacitly uses a set of linguistic rules. Knowledge-based fuzzy tuning has its roots in a knowledge base of such rules. In a typical situation the human expert will observe a particular condition and take one or more tuning actions before turning attention to another condition. In this sense the conditions are considered sequentially and repetitively. For example, in tuning a PID servo, if an oscillatory (instability) condition is noticed, a first action would be to increase the rate feedback parameter and decrease the loop gain. The expert may subsequently take appropriate tuning actions for a speed of response problem. This type of sequential observation of condition variables will lead directly to a dissociation of rules. Furthermore, in practice, the observation-action cycles are incremental in nature. For example, if a strong instability is observed, a large tuning adjustment would be made, and subsequently if the condition becomes moderate, the tuning actions would be toned down. Under these conditions, the linguistic tuning relation between a particular attribute (E_s) and the corresponding action (C_q) may be viewed as a sequence of nonotonically incrementing rules; specifically:

$$R_{s,q} : \underset{i=0}{\overset{m-1}{\text{ELSE}}} [\text{IF } E(i) \text{THEN } C(i)] \tag{8}$$

The subscripts s and q have been dropped in the rules given by Equation (8), for convenience. The membership function of $R_{s,q}$ is formed by the *max-min* composition

$$\mu_{R_{s,q}}(e,c) = \underset{i=0}{\overset{m-1}{max}} min[\mu_{E(i)}(e), \mu_{C(i)}(c)] \tag{9}$$

The complete rule base is represented by

$$R = \underset{s,q}{U} R_{s,q} \tag{10}$$

This rule dissociation will result in a substantial reduction of the computational requirement in terms of *min* and *max* operations, and is discussed later in this chapter.

Resolution Relationships

The monotonically incrementing characteristic that is exhibited in the rule base [Equation (8)] will lead to a quantification of fuzzy resolution. In connection with the rule base [Equation (2)], it was mentioned previously that the total number of possible fuzzy states for a fuzzy variable will represent its fuzzy resolution. For example,

E_s and C_q in Equation (2) can assume, respectively, m and r fuzzy states. Unfortunately, in Equation (2) the resolutions are fixed at the outset in the problem definition, the rule base must be modified, and the entire problem must be reformulated if one wishes to change resolutions. The incremental rule base given by Equation (8) is attractive in this respect because here the fuzzy resolution is linked to the index i and can be refined arbitrarily and varied during operation of the tuner without having to reformulate the problem. The concept of resolution relationships is introduced here to formalize fuzzy resolution and to further strengthen the analytical framework of the fuzzy tuning problem.

Suppose that the membership function $\mu_{E(i)}(e)$ has a modal value $\bar{e}_i$ and similarly $\mu_{C(i)}(c)$ has a modal value $\bar{c}_i$. These membership functions are assumed to be unimodal. Typically the modal value is the value of the variable at the peak membership grade, and this usage is particularly convenient when triangular membership functions are used as in the examples of this chapter. An alternative modal value (e.g., the centroid of the membership function) may be used if its use can be justified. This may lead to increased computational complexity, but the concepts presented here will be still applicable. Generally, the resolution relationships for $E(i)$ and $C(i)$ are given by

$$\bar{e}_i = r_e(i) \quad i = 0, 1, \ldots, m-1 \tag{11}$$

$$\bar{c}_i = r_c(i) \quad i = 0, 1, \ldots, m-1 \tag{12}$$

in which r_e are r_c monotonic functions in the index i, and these functions will determine the fuzzy resolution of the corresponding variables. Specifically, one could emphasize various tuning levels by appropriately selecting the resolution functions. For example, if linear relations are used, all tuning actions are uniformly emphasized, and if exponential functions are used, large changes are more heavily emphasized than small changes. Polynomial resolution relationships of the form

$$\bar{e}_i = a_e i^{m_e} \quad i = 0, 1, \ldots, m-1 \tag{13}$$

$$\bar{c}_i = a_c(i^{m_c} - \beta) \quad i = 0, 1, \ldots, m-1 \tag{14}$$

are normally appropriate. Note that the coefficients a_e and a_c are positive parameters representing intensities, and the exponents m_e and m_c are positive (typically integer) parameters representing the degree of nonlinearity. The parameter β provides an offset which will tend to reduce the magnitudes of the tuning parameters under quiescent conditions. Care must be exercised in selecting the value of this parameter because unsteady behavior may result with large values. In this manner fuzzy resolution can be conveniently modified on line during operation of the knowledge-based tuner without having to reformulate the entire problem. In practice lower and upper bounds must be introduced to the sequences in Equations (13) and (14) in order to incorporate physical constraints or stability bounds of the tuning parameters. These bounds may be known through experience.

Resolution relationships can be introduced even if complete rule dissociation is not carried out. In this case the resolution relationships are multivariable and will relate the modal values of several attribute variables to the modal value of an action variable; for example,

$$\bar{c}_{i,j,\cdots,k} = f_r(\bar{e}_i, \bar{e}_j, \cdots, \bar{e}_k) \qquad (15)$$

A resolution relationship provides an analytical representation of fuzzy states and fuzzy resolution of the condition and action variables that are related by a fuzzy rule. It gives the progression of the modal values which characterize the associated unimodal membership functions. In this latter sense it affords some form of "analytical interpolation" between alternative fuzzy states in an appropriately partitioned rule base.

To summarize, *fuzzy resolution* of a (fuzzy) variable is determined by the number of possible fuzzy states the particular variable is allowed to assume within an analytical formulation of the control (tuning) problem. The larger the number of states, the finer the resolution. The fuzzy resolution can be analytically represented by the "increment" of a corresponding index. A *resolution relationship* provides an analytical means of incorporating fuzzy resolution into a tuning scheme. Specifically, a fuzzy state is represented by an indexed and unimodal membership function; the membership function has a modal value which is also indexed; and an analytical relationship between the indexed modal values of a condition variable and a corresponding action variable provides a resolution relationship. Resolution relationships that are nonlinear and complex may be conveniently expressed in terms of resolution functions. A *resolution function* is a function of the indexing variable, which represents the modal values of a membership function.

Tuning Inference

Suppose that the membership function of the dissociated rule base is determined as given by Equation (9). The compositional rule of inference must be applied to this rule base, with a crisp measurement $\hat{e}_s$ of the corresponding response attribute E_s. The measurement may be represented as a fuzzy singleton $s(e - \hat{e}_s)$:

$$\mu_{\hat{E}_s}(e) \triangleq s(e - \hat{e}_s) = 1 \quad \text{for } e = \hat{e}_s$$
$$= 0 \text{ elsewhere,} \qquad (16)$$

which is essentially the first integral of the Dirac delta function. The compositional rule of inference, as given by Equation (6), can be applied here; thus

$$\mu_{\hat{C}_q}(c) = \sup_e \min[\mu_{\hat{E}_s}(e), \mu_{R_{s,q}}(e,c)]$$
$$= \sup_e \min[s(e - \hat{e}_s), \mu_{R_{s,q}}(e,c)]$$

which, in view of Equation (16), becomes

$$\mu_{\hat{C}_q}(c_q) = \mu_{R_{s,q}}(\hat{e}, c) \qquad (17)$$

Finally, a crisp tuning action $\hat{c}_q$ may be computed according to Equation (7). Note that the tuning actions are incremental and not continuous, and an increment is governed by the fuzzy resolution that is employed, as discussed in the previous subsection.

Stability Considerations

Let us again consider the system model given by Equation (1). A conventional parameter-adaptive controller for this system has the general form

$$\dot{\mathbf{c}} = \mathbf{g}(\mathbf{e}, \mathbf{c}) \quad \text{in which } \mathbf{g} : \Re^n \times \Re^p \to \Re^p \qquad (18)$$

In this case the dynamic characteristics, including stability of the overall adaptive control system, is determined in an $\Re^{n+p}$ state space and the dynamic behavior of this system is altered in a nonlinear and coupled manner. Alternatively, if the adaptive law is algebraic rather than differential, as given by

$$\mathbf{g}(\mathbf{e}, \mathbf{c}) = \mathbf{0} \qquad (19)$$

the stability would be determined in the original $\Re^n$ state space.

Similar observations can be made regarding the fuzzy tuning system. However, because in the case of interest to us here the condition attributes that are observed are not simply the error state vector, but rather some characteristics that are of direct relevance in manual tuning, as established through expertise of the control engineer (intelligent preprocessing), we must include an additional transformation in considering system stability. Once the control system is tuned, its behavior (including stability) is completely governed by Equation (1), with tuned values for **c**. Hence, dynamics during the tuning process only need be considered.

Suppose that $S_{\hat{e}}$ is the set of possible process attribute observations and that $S_{c'}$ is the corresponding set of tuning actions. Knowledge-based fuzzy tuning then corresponds to the mapping $S_{\hat{e}} \to S_{\hat{c}}$. Also, suppose that $S_{e'}$ is the set of error states which produce the attribute set $S_{\hat{e}}$. The mapping $S_{e'} \to S_{\hat{e}}$ is also needed in order to establish an analogy between the parameter-adaptive control problem and the fuzzy tuning problem. The overall mapping $S_{e'} \to S_{c'}$ will determine the dynamics of the knowledge-based tuning system. Specifically, if this mapping is algebraic, the stability of the overall system will be determined by the original system (Equation (1)), subject to this algebraic relation. This is the case if both $S_{e'} \to S_{\hat{e}}$ and $S_{\hat{e}} \to S_{\hat{c}}$ are algebraic.

The parameter vector **c** is interpreted as tuning increments rather than absolute values of the tuning parameters. Then, in the dissociated problem, the situation of particular interest here, it is clear in view of Equations (7) and (17) that the mapping $S_{\hat{e}} \to S_{\hat{c}}$ can be considered algebraic, even though highly nonlinear, at least when discrete trapezoidal integration is used in Equation (7). Validity of the assumption of an algebraic mapping for $S_{c'} \to S_{\hat{e}}$ will depend largely on the nature of the performance attributes **ê** employed in the tuning rule base, and must be determined on a case-by-case basis. For instance, suppose that *percentage overshoot, error divergence, steady-state error*, and *speed of response* are the four attributes used in the rule base, as considered in the example of Chapter 4. The first three attributes can be taken as algebraically related to the error state, at least during specific time periods, as determined by additional logic. If the rate of change of error is included as a state variable, one can use an algebraic expression of this variable to represent the speed of response. In this particular case the mapping $S_{e'} \to S_{\hat{e}}$ can be assumed to be algebraic. Then, stability of the overall knowledge-based control system will be governed by the system given by Equation (1) in $\Re^n$ subject to the algebraic relationship between **e** and **c**. Because the tuning bandwidth (the highest frequency at which the parameters are tuned) is in practice an order of magnitude smaller than the control bandwidth (the highest frequency at which the control signals are updated), the stability of the overall system can be determined from the stability regions of **c** in Equation (1). These regions may be incorporated as lower and upper bounds for the resolution sequences of Equations (13) and (14). Alternatively, other common methods of studying the stability of an adaptive control system (Narendra and Annaswamy, 1989) may be applied.

An alternative, system-level interpretation of stability is possible for fuzzy logic control systems. In this case the stability of the decision maker itself is considered. In fuzzy logic terms, then, what is addressed is the stability of successive application of the compositional rule of inference. This concept of stability of a fuzzy logic control system is further addressed in Appendix A.

■ COMPUTATIONAL EFFICIENCY

Computational requirement for knowledge-based tuning is addressed here in terms of the number of maximum (*max*) and minimum (*min*) operations needed to generate the tuning rule base and to obtain a tuning inference. The dissociated problem, as presented in the Analytical Framework section, is compared to the direct problem of the Theoretical Background section on this basis.

Direct Formulation

Consider the rule base R given by Equation (4). Suppose that N discrete points are used to digitally represent any of the membership functions $\mu_{E_s}(e)$ and $\mu_{C_q}(c)$. The rule base membership function $\mu_R(\mathbf{e}, \mathbf{c})$ is a hypertable of $N \times N \times \cdots \times N$ ($n + p$ times) values, given by $\Re^{n+p} \to [0,1]$. For each point in $\mu_R(\mathbf{e}, \mathbf{c})$ and for each rule one would need a maximum of $n + p - 1$ *min*. Now, because each E_s can assume m different states, the rule base will be limited to a maximum of m^n rules. Consequently, one would need $(n + p - 1)$ m^n *min* and $(m^n - 1)$ *max* to compute one point in the hypertable of μ_R. Because there are a maximum of N^{n+p} points in the table a maximum total of $(n + p - 1) m^n N^{n+p}$ *min* and $(m^n - 1) N^{n+p}$ *max* will be needed to generate the discrete rule base μ_R.

Next, consider the compositional rule of inference as expressed by Equation (6). For each tuning parameter q and for each discrete point c_q of the tuning membership function $\mu_{\tilde{C}_q}$ there are $N \times N \times \cdots \times N$ (n times) combinations of values for $\boldsymbol{\mu}_{\tilde{E}}(\mathbf{e})$ disregarding that fuzzy singletons might be present. Each such combination determines a unique value for $\mu_{R_q}(\mathbf{e}, c_q)$, and together will require n *min*. Hence, the entire set of N^n combinations, a total of nN^n *min* and $(N^n - 1)$ *max* will be needed to evaluate one point of the tuning membership function. Consequently, for all N points in a tuning membership function and for all p such membership functions one needs $pn N^{n+1}$ *min* and $pn (N^{n+1} - 1)$ *max*. In summary,

$$\text{Rule - base total} = (n + p - 1)m^n N^{n+p} \, min + (m^n - 1)N^{n+p} \, max$$
$$\text{Inference total} = pnN^{n+1} \, min + pN(N^n - 1) \, max$$
(20)

These estimates are conservative because they represent the maximum number of operations. Various economizations are possible by eliminating redundant operations due to the presence of the special membership grades of 0 and 1. Further economization may be achieved through memory-intensive approaches. For example, consider the compositional rule of inference [Equation (6)] again. Suppose that the *min* computations are made according to the following steps:

First compute $min[\mu_{\hat{E}_2}, \cdots, \mu_{\hat{E}_n}] \triangleq \mu_{min} \Rightarrow n-2$ *min* for one point in the sub hypertable

Next compute $min[\mu_{\hat{E}_1}, \mu_{min}] \triangleq \mu^* \Rightarrow 1$ *min* for each discrete value in $\mu_{\hat{E}_1}$

Finally compute $min[\mu^*, \mu_{C_q}]_{q,N} \Rightarrow pN$ *min* for all N discrete values in μ_{C_q} and for all p parameters

Total for all discrete values in $\mu_{\hat{E}_1} = n - 2 + (1 + pN)N$ *min*

There are N^{n-1} discrete value combinations for $\{\hat{E}_2, \cdots, \hat{E}_n\}$ Hence

Total $= [n - 2 + (1 + pN)N] \times N^{n-1}$ *min* $= pN^{n+1} + N^n + (n-2)N^{n-1}$ *min*

Note that an improvement approximately by a factor of n has been achieved through this approach.

As can be seen from Equation (20) the computing time is significantly affected by the fuzzy resolution m of e_s, but the fuzzy resolution r of c_q does not appear in the equations. This is because (1) the maximum number of possible rules (m^n) in the rule base is determined by the number of condition variables and their fuzzy resolution m, and hence the computation of the rule base is not affected by the fuzzy resolution r of c_q, and (2) the computation of the inference concerns only the rule base and the membership functions of **e**; i.e., the fuzzy resolution r does not participate directly in the inference procedure. In summary, fuzzy resolution of c_q does not affect the defuzzification process. Compared to the time for computing the rule base and the inference, however, the time taken for the fuzzification and defuzzification processes is negligible. Accordingly, the analysis for this computation is omitted here.

To graphically illustrate how the fuzzy resolution m affects the total computing time, the relation represented by Equation (2) is plotted in Figure 2 for the case $N = 100$, $n = 5$, $p = 4$, and m varies from 1 to 15. We assume here that it takes the same computing time unit for a *min* operation as for a *max* operation. As can be seen from the figure, the computing time increases exponentially with the fuzzy resolution m of e_s.

Accuracy vs. Fuzzy Resolution

Appropriate use of the level of fuzzy resolution is important in a knowledge-based fuzzy control system. Generally speaking, the higher

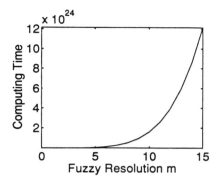

Figure 2 An illustration of the variation of computational time with fuzzy resolution.

the fuzzy resolution that the variables can assume, the more accurate control a system can achieve. However, as the resolution increases, generally the computational time will also increase. This may affect the performance of the control system, for example, with regard to the control bandwidth, or it may require more expensive hardware. More importantly, the stability of the system may be affected by the increased fuzzy resolution. For example, if a low fuzzy resolution is used in the system, the number of rules that define different input conditions would be limited, and there is a good possibility that no rules will define certain input conditions or states. Such undefined or ambiguous situations can lower the efficiency of the fuzzy controller. This problem can be avoided or partially resolved by increasing the fuzzy resolution. The accuracy of the system will improve consequently, but the operating bandwidth will decrease due to increased computational effort. On the other hand, only a coarse tuning action may be generated by a system with low fuzzy resolution. Hence, it is possible that the system will exhibit a hunting response about the desired operating point.

To illustrate this point further, assume that a series of "well-behaved" input-output data is available, which could be generated, e.g., by a human or well-tuned automatic controller. Fuzzy associative memory (FAM) rules (Kosko, 1992) can be obtained using an adaptive vector quantization (AVQ) algorithm with differential competitive learning (DCL). In control applications, given the crisp values of both input and output variables, the DCL-AVQ method can be used to adaptively cluster these data vectors to estimate the underlying control surface (continuous rules) that generated them. The product space is then partitioned into FAM cells, each one representing a single FAM rule (Kosko, 1992). The maximum number of FAM cells is determined by the product mp, as defined in Equation (2), and the number of FAM cells will determine the smoothness of the control surface. A smoother control surface implies that given the data for input conditions, finer tuning actions could be generated; i.e., a more accurate relationship between the performance conditions and the tuning actions could be established.

On the other hand, as mentioned earlier, the processing speed is affected directly by the fuzzy resolution, and accordingly, the heavy computational burden resulting from a high fuzzy resolution can slow down the response, thereby reducing the operating bandwidth. In order to analyze the effect of fuzzy resolution on the accuracy suppose that the computational efficiency is not affected by the fuzzy resolution. In practical implementations this could be achieved by developing a crisp decision table off-line and using this table in a table look-up mode to determine a crisp control action during operation (de Silva and MacFarlane, 1989). However, if a decision table is used, the number of discretization levels would be limited by the memory capacity of the controller; only a coarse control surface could be achieved with a few discretization levels.

Dissociated Formulation

In the dissociated problem, the rule base is partitioned as in Equation (8) so that each partition associates one performance attribute with a corresponding tuning action. (This does not exclude the case of having many tuning actions for a given performance condition; the actions are taken sequentially.) Hence, for each partition we have $n = 1$. Also, the fuzzy resolution is now determined by the range of i and this is taken to be m, as given in Equation (8). As before, the membership functions $\mu_{E(i)}(e)$ and $\mu_{C(i)}(c)$ are digitally represented by N discrete points. It follows that there are N^2 discrete points in $\mu_{R_{s,q}}(e, c)$ which is defined according to $\Re^2 \to [0,1]$. For each of these discrete points (e, c), it is clear from Equation 9 that one needs m *min* and $m - 1$ *max*. It follows that for all N^2 points and for all p tuning parameters:

Rule base total = pmN^2 *min* + $p(m - 1)N^2$ *max* for each performance attribute

For the n performance attributes this total must be multiplied by n. By comparing this total with the rule base total of *min* and *max* for the direct computation it is evident that the rule base computations increase exponentially with the number of performance attributes and tuning parameters in the direct formulation, whereas it increases only linearly in the dissociated formulation. This represents a substantial reduction in the computational burden.

The foregoing analysis assumes that the support set of $\mu_{R_{s,q}}(e, c)$ has N^2 points. In fact, in practice much better resolution is achieved in $\mu_{R_{s,q}}(e, c)$ through suitable analytical representations of the

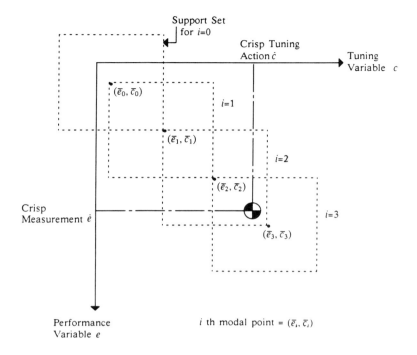

Figure 3 The support set of a dissociated rule.

membership functions, and furthermore, obviously redundant *max* operations can be avoided. This may be explained using Figure 3 and is illustrated further in the example at the end of the chapter.

The fuzzy overlap between two adjoining fuzzy states (e.g., $i - 1$ and i) is assumed to be rectangular and extends only from one modal point to the adjoining one, as illustrated in Figure 3. This assumption is justifiable because if the overlap extended significantly beyond this region, the corresponding fuzzy states could be combined to give a fewer number of states. Because each of the two indexed fuzzy variables $E(i)$ and $C(i)$ has a support set of N discrete points, it is clear that $min[\mu_{E(i)}(e), \mu_{C(i)}(c)]$ in Equation (9) has a support set of N^2 discrete points. Then, with a fuzzy overlap as shown in Figure 3 it is clear that only $N^2/4$ *max* operations will be needed for each region of overlap. Consequently, the total number of *max* needed for computing Equation (9) would be $(m - 1)N^2/4$ for each performance attribute and tuning parameter or $np(m - 1)N^2/4$ for all n attributes and p parameters. This results in a further reduction of the *max* operations by a factor of 4, while improving the resolution of the overall rule base. Note that when the membership function of the dissociated rule base is available in this manner, in view of Equation (17) and as graphically represented in Figure 3, no further *min* and *max* operations will be needed to compute a tuning inference by applying the compositional rule. Only a defuzzification computation, as given by Equation (7), will be needed.

■ DYNAMIC SWITCHING OF FUZZY RESOLUTION

Theoretically, a fuzzy variable may assume any finite number of fuzzy states. (When there is an infinite number of fuzzy states, a fuzzy variable behaves like a crisp variable.) In practice, however, just three to nine fuzzy states are typically adequate for a system variable. For example, when an operator describes a control action, the descriptors such as negative large (NL), negative small (NS), no change (NC), positive small (PS), and positive large (PL) may be used. Therefore, fuzzy resolution represents an approximate characterization of the level or state of a variable. To be precise, the following sections define fuzzy resolution of a variable through its membership functions. This topic is revisited in Chapter 8.

Here, we look at only a few specific situations that are often encountered in practical applications. We assume that each fuzzy state of a variable has a unique modal point (typically, having a peak membership grade) in the membership function; then the number of such modal points in the membership functions (within its support set) corresponds to the number of fuzzy states of the variable (within the knowledge base). Let us further assume that the membership functions are symmetrical and they uniformly partition the support set. Then the intermodal spacing of the membership may be used as a measure of the *fuzzy resolution* of the variable. Mathematically, this can be described as follows.

Let X be a universe of discourse, x be its generic element, and $\mu_A(x):\Re \to [0,1]$ be the membership function of a fuzzy set A. Then an appropriate quantitative definition for fuzzy resolution would be

$$r_j = \frac{w_s}{w_m} \qquad (21)$$

where w_s represents the width of the support set of A and w_m is the intemodal spacing. If the membership functions have trapezoidal shapes, the intercentroidal spacing (the distance between the centroids of two fuzzy sets) should be used in place of the intermodal spacing.

Another consideration of importance is the *fuzziness* of a fuzzy descriptor. The degree of fuzziness is assumed to express, on a global level, the difficulty of deciding which elements belong and which do not belong to a given fuzzy set. There are several quantitative measures available for this purpose in the literature (Dubois and Prade, 1980; Terano et al., 1992). Here, the degree of fuzziness of a set A is measured by the closeness of its elements to the membership grade of 0.5. This is justified because the membership of an element x is most fuzzy when the membership grade is 0.5, i.e., when $\mu_A(x) = 0.5$ (de Silva, 1993). Mathematically, then, the fuzziness can be defined as

$$f_A = \int_{x \in X} \left| \mu_A(x) - \mu_{A_{1/2}}(x) \right| dx \tag{22}$$

Here, $\mu_{A_{1/2}}$ is the $1/2$-cut of A, as presented in Chapter 3. Generally, α-cut of a fuzzy set A is the crisp set formed by elements of A whose membeship grade is greater than or equal to a given value α. This is denoted by A_α. Specifically,

$$A_\alpha = \{x \in X | \mu_A(x) \geq \alpha\} \tag{23}$$

in which X is the universe to which A belongs. With this definition, it can be shown that in the context of a hierarchical fuzzy tuning system, a higher layer of the control hierarchy usually possesses a higher degree of fuzziness in its knowledge base (de Silva, 1991a; 1993). This is because high-level decision making requires high-level information, which is typically generated through a signal preprocessor. The preprocessed information is generally fuzzier and has a lower resolution. This aspect is further discussed in Chapter 8.

The concepts of *resolution relationship* and dynamic switching of fuzzy resolution (referred to as *resolution switching*) are used in the following sections. The concept of resolution relationship was first introduced by de Silva (1991b), and has been discussed previously in the present Chapter. Suppose that the membership functions of all fuzzy sets considered are unimodal and the modal value is the value of the variable at the peak membership grade. Then, as discussed previously, the resolution relationship can be established for the condition variables and action variables; i.e., the variables for measured conditions and tuning actions. These are given, respectively, by Equations (13) and (14).

Resolution switching can be considered as a particular case of nonlinear resolution relationship (Wu and de Silva, submitted). We say that there is resolution switching if the resolution relationship can be represented as

$$\bar{x}_i = \begin{cases} \alpha_1 i \text{ for } 0 \leq i < m_1 \\ \alpha_2 i \text{ for } m_1 \leq i < m_2 \\ \quad \cdots \quad \cdots \\ \alpha_j i \text{ for } m-2 \leq i < m-1 \end{cases} \quad (24)$$

where $\alpha_1, \alpha_2, \ldots, \alpha_j$ are coefficients which characterize different fuzzy resolution levels, $m_1, m_2, \ldots,$ are suitably defined threshold values within the support set, and $\bar{x}_i$ represents the modal value of either a condition variable or an action variable. In this definition linear resolution relationships are employed between two switching points. It can be seen later in the chapter that the use of resolution switching is more convenient than using a continuous or monotonic nonlinear resolution relationship, as defined in Equations (13) and (14), especially in establishing an analytical tuning criterion for a fuzzy control system.

Resolution Switching

From the discussions in the previous section it is clear that a control system with low fuzzy resolution has the advantages of fast processing, high control bandwidth, and easy management of the rule base. However, only a coarse control action would be generated which might not be accurate enough for the purpose of fine control and tuning. On the contrary, a control system with high fuzzy resolution is capable of generating more accurate control and tuning actions. However, it is computationally expensive and its rule base is difficult to manage. This suggests that if a resolution switching capability is incorporated into the fuzzy control system, the performance of the system could be improved, because in this case a low fuzzy resolution could be used for coarse control and tuning. Once the system is in the neighborhood of the steady state, a finer fuzzy resolution could be used for control and fine tuning.

Fuzzy Tuning System with Fixed Resolution

Several versions of a fuzzy system for tuning the parameters of a PID controller have been reported, as discussed in this book and also by He (1992). It may consist of a three level hierarchy: the hard (crisp) control level, the servo expert level, and the knowledge-based fuzzy tuning level. The lowest level has a direct PID control loop with position feedback. This will guarantee that high-bandwidth control can be achieved because the time-consuming knowledge-based reasoning is not directly involved in this loop. The second level may contain a servo expert, which evaluates the preprocessed system response and generates performance indices of the response attributes. These performance indices are then fed into the knowledge-based tuning layer, which will occupy the highest level of the control hierarchy. Compositional rule of inference and defuzzification are performed in this highest layer and a decision table is generated and used to update the PID parameters of the servo controller, as discussed in Chapters 4 and 6.

When a fixed fuzzy resolution is used, typically the rule base shown by Equation (2) is directly implemented. Thus, every time the fuzzy resolution changes, the rule base must be modified and the entire problem must be reformulated following the procedures of fuzzification, application of the compositional rule of inference, and defuzzification. This makes it difficult to switch the fuzzy resolution of the system on-line during operation of the system. The *resolution relationship*, as defined in the previous section, is very useful in this context.

This chapter uses the concept of resolution relationship to derive an analytical formulation for the tuning system incorporated with resolution switching capability, by following the procedures introduced in de Silva (1991b). In this approach the fuzzy resolution is represented by an index, and may be refined arbitrarily and varied during operation of the tuner, without having to reformulate and recompile the problem. This approach is certainly advantageous in terms of the convenience in analyzing and implementing the knowledge-based tuner, especially when studying the effect of the fuzzy resolutions of the condition and action variables. It will also make the implementation of the resolution switching fuzzy system much easier, as shown in the following section.

Fuzzy Tuning with Resolution Switching

In our implementation of a fuzzy tuning system each fuzzy rule is used in mapping several condition variables (E) to one or more action variables (fuzzy tuning parameters C). Here, the membership functions of the action variables C are obtained by Equation (6). With the introduction of an *incremental rule base* and a *resolution relationship*, it is possible to establish an analytical tuning criterion for each pair of tuning condition-action attributes (de Silva, 1991b). Two steps are followed to formulate such tuning criteria in our implementation. Details are given in Wu and de Silva (submitted).

First, the rule base is dissociated for the action part by considering the tuning parameters sequentially. For example, in a PID controller the tuning actions can be arranged in the order D-P-I. Next, for the condition part, the *decomposition theorem* (Wang and Vachtsevanos, 1990) is applied to decouple the rule base; thus

$$\mu_{C_q}(c_q) = \{sup\ min[\mu_{E_1}(e_1), \mu_R(e_1, c_q)]\} \wedge$$

$$\{sup\ min[\mu_{E_2}(e_2), \mu_R(e_2, c_q)]\} \wedge \cdots \qquad (25)$$

$$\{sup\ min[\mu_{E_n}(e_n), \mu_R(e_n, c_q)]\} \qquad q = 1, 2, \cdots, p$$

Next, we assume that the membership functions used for all fuzzy sets are of the form

$$\mu_{X_i}(x) = \begin{cases} max\left(0, \dfrac{x - \bar{x}_{i-1}}{\bar{x}_i - \bar{x}_{i-1}}\right) & \text{for } x \leq \bar{x}_i \\ max\left(0, \dfrac{\bar{x}_{i+1} - x}{\bar{x}_{i+1} - \bar{x}_i}\right) & \text{for } x \geq \bar{x}_i \end{cases} \qquad (26)$$

where X represents either a condition variable or an action variable. Note that $\bar{x}_i$ is the modal value, and $\bar{x}_{i+1} - \bar{x}_{i-1}$ is the spread of the fuzzy set.

Armed with these tools, the entire resolution switching system can be formulated by the following steps:

1. *Selection of condition variables and action variables:* Condition variables and action variables can be chosen according to the requirements of the specific application and the architecture of the fuzzy control/tuning system. As shown in the context of experimental studies, if step signals are applied as the test inputs in tuning, the parameters related to the step response may be used as condition variables.

2. *Determination of resolution switching frequency:* Suppose that the number of fuzzy states which a condition or an action variable can assume is m. (This resolution could be changed during on-line tuning.) The resolution-switching frequency is determined according to the needs of the practical application. For example, it may be appropriate to use just one threshold of switching for some applications, and more for some others. The threshold values at which the switchings occur may be chosen, for instance, to equally divide the range of a condition attribute. Alternatively, they could divide the range in a nonuniform manner, for example, wider when it is far away from the set point and narrower when it is closer to the set point.

3. *Formulation of nondimensional tuning criterion:* The tuning criterion is formulated which can be used by every condition-action pair (e.g., $c_1 \to K_p$). The procedure for formulating such an analytical tuning criterion is similar to the one proposed in de Silva (1991b); specifically, the membership function for each tuning criterion at index i is given by

$$\mu_{R_i} = max(0, min[e_i^*, c_i^*]) \qquad (27)$$

where

$$e_i^* = \begin{cases} \dfrac{e - \bar{e}_{i-1}}{\bar{e}_i - \bar{e}_{i-1}} & \text{for } e \leq \bar{e}_i \\ \dfrac{\bar{e}_{i+1} - e}{\bar{e}_{i+1} - \bar{e}_i} & \text{for } e > \bar{e}_i \end{cases} \qquad (28)$$

$$c_i^* = \begin{cases} \dfrac{c - \bar{c}_{i-1}}{\bar{c}_i - \bar{c}_{i-1}} & \text{for } c \leq \bar{c}_i \\ \dfrac{\bar{c}_{i+1} - c}{\bar{c}_{i+1} - \bar{c}_i} & \text{for } c > \bar{c}_i \end{cases} \qquad (29)$$

It should be noted that by representing a crisp measurement $\hat{e}_s$ of the corresponding condition attribute E_s as a fuzzy singleton and considering the fact that the rule base is dissociated, the following result is obtained:

$$\mu_{C_q}(c) = \mu_{R_{s,q}}(\hat{e}_s, c) \qquad (30)$$

for the action variable C_q. This result is applied to each index i of Equation (28). Then, to obtain the fuzzy output for the overall action variable C_q, a *max* operation is applied over the entire range of the index i of resolution states. The above formulation is applicable when a linear resolution relationship is used. It must be modified when adaptive resolution is incorporated. Two approaches can be followed to derive the analytical tuning criterion in this case. The first one uses linear resolution relationship within each threshold interval, as shown in Figure 4(a). This corresponds to resolution switching. The second approach uses a nonlinear resolution relationship, as shown in Figure 4(b). Here, the resolution is changed (or "adapted") continuously. Formulation of an analytical tuning criterion using a nonlinear resolution relationship is somewhat more complex than using a linear one. Therefore, for simplicity, the first approach is explored here and used in the experiments. In this case the tuning criterion formulated above can still be used provided that the resolution relationship is modified as follows:

$$\bar{x}_i = i\tilde{x} \qquad (31)$$

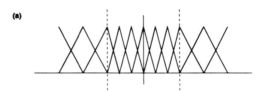

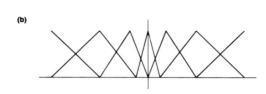

Figure 4 Two types of membership functions: (a) for dynamic resolution switching, and (b) with a general nonlinear resolution relationship.

$$\tilde{x} = \begin{cases} \overline{x}_f & \text{for } \overline{x} < \overline{T}_s \\ k\overline{x}_f & \text{for } \overline{x} \geq \overline{T}_s \end{cases} \quad (32)$$

where $\overline{x}_i$ represents the scaling factors for the corresponding condition and action variables, and $\overline{T}_s$ represents a vector of threshold values at which the dynamic switching of fuzzy resolution occurs.

4. *Coupling of fuzzy outputs:* Because the action part was dissociated by using the decomposition theorem for the purpose of formulating the analytical tuning criterion, the fuzzy outputs generated should be recoupled by applying the operation given by Equation (25). Specifically, the *min* operation is applied to the membership functions of all the fuzzy outputs that were obtained in the previous step.

5. *Defuzzification to generate crisp tuning actions:* Fuzzy action variables obtained as described in the previous steps, must be defuzzified into crisp quantities. This can be accomplished by the centroidal defuzzification method.

Experiments and Results

A set of experiments is conducted on an industrial robot retrofitted with a low-level programmable digital controller. The objective is to make the joints of the robot follow a specified trajectory and to use the results to study the tracking accuracy. On-line tuning is performed using error and error rate as the observed responses, while the trajectories are being tracked.

Most industrial robots use simple PID-type controllers. Also, usually their low-level control loops cannot be accessed or programmed by the user. In view of the presence of factors such as nonlinearities,

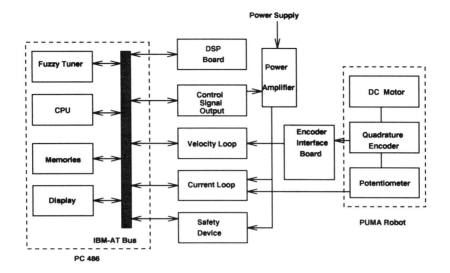

Figure 5 The schematic diagram of the retrofitted robotic testbed.

dynamic coupling, time variance, unknown loads, and external disturbances, it is often desirable to employ more advanced control schemes such as adaptive control or intelligent control (see Chapter 1). To implement and test various advanced control algorithms, an industrial robot (PUMA 560) has been retrofitted in our laboratory with a software-programmable, low-level controller. Figure 5 shows the schematic diagram of the retrofitted robotic testbed. The set of fuzzy self-tuning control experiments is conducted using this experimental system.

The testbed consists of a PUMA 560 industrial robot and a custom-built controller centered around a 486 PC in conjunction with a digital signal processing (DSP) board for high-speed computation. The Intel80486 CPU on the PC is the central controller of the system. It runs the operating system and performs system initialization, self-diagnosis, and input-output control. The computation of the control algorithm is performed by the DSP board, which is capable of executing 25 million floating point operations per second. Other modules of the controller include the joint position and velocity loops, the current loop, the encoder data acquisition board, the control signal output channel, the interrupt control circuits, the power amplifiers, the safety device and the power supply module. The software module is implemented on the 486 PC enhanced with the DSP board.

The criteria used to describe the performance of a step response are not suitable for tracking a special trajectory, and the error together with the error rate are used as the condition variables in this set of experiments. The resolution switching fuzzy tuning system implemented for this experiment is based on a fuzzy self-tuning PID controller given in He (1992). The major steps are outlined below.

The fuzzy self-tuning PID controller consists of a standard PID controller and a fuzzy tuning mechanism for on-line adaptation of the PID parameters. The PID controller has the following standard form:

$$u(t) = K_c[e(t) + T_d \frac{de(t)}{dt} + \frac{1}{T_i} \int e(t)dt] \tag{33}$$

where K_c, T_d, and T_i, respectively, are the proportional gain, the derivative time constant, and the integral time constant of the controller. These three parameters, which may be parameterized by a single parameter α, are updated using the following equations (He, 1992):

$$K_c = 1.2\alpha\, k_u \tag{34}$$

$$T_i = 0.75 \frac{1}{1+\alpha} \tau_u \tag{35}$$

$$T_d = 0.25 T_i \tag{36}$$

$$\alpha(t+1) = \begin{cases} \alpha(t) + \gamma h(t)(1 - \alpha(t)) & \text{for } \alpha(t) > 0.5 \\ \alpha(t) + \gamma h(t)\alpha(t) & \text{for } \alpha(t) \leq 0.5 \end{cases} \tag{37}$$

where k_u and τ_u are, respectively, the ultimate gain and the ultimate period of the underlying process, γ is a positive constant used to modify the convergence rate of the updating formulas, and $h(i)$ denotes the action variable. Here, k_u and τ_u are determined by using the relay feedback method (Astrom and Hagglund, 1989). The mapping from the fuzzy condition variables E and E' to the fuzzy action variable H is achieved by using the analytical tuning criterion, which is established and coded into the routines by following the procedures described previously.

In this experiment the second and the third joints of the robot are programmed to track sinusoids with periods of 1.5 and 3 s, respectively. The higher frequency for the third joint was intended to present a significant disturbance to the second joint. Figure 6 (top

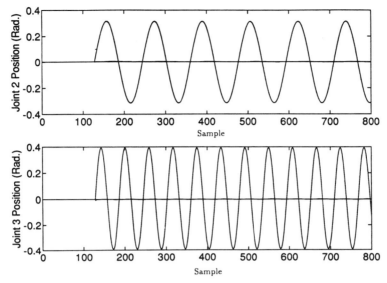

Figure 6 Joint responses of a Puma 560 robot under tuning control with resolution switching. (Top) Second joint. (Bottom) Third joint.

and bottom) shows the responses of the second joint and third joint, respectively, generated using the self-tuning control system with a resolution switching mechanism. An equivalent fuzzy resolution of 5 is used here. The figures show that both joints can track the given trajectories with high accuracy.

An alternative approach to using the characteristics of the trajectory error (de Silva and MacFarlane, 1989) in tuning the parameters of the low-level controller of the robot is to actually apply step perturbations to the input signal, under quiescent conditions, and to use the corresponding deviations of the response to ascertain the performance of the control system. In this manner the controller could be tuned at specific points of the robot trajectory, using the rule base developed on the basis of step response, without adversely affecting the overall response of the robot.

ILLUSTRATIVE EXAMPLE

To illustrate much of the analytical concepts developed in this chapter, consider the example of a nonminimum phase plant with transport delay, as shown in Figure 7. The plant is controlled directly by a PID loop. Knowledge-based tuning of the PID parameters, the proportional gain K_p, integral gain K_i, and the derivative

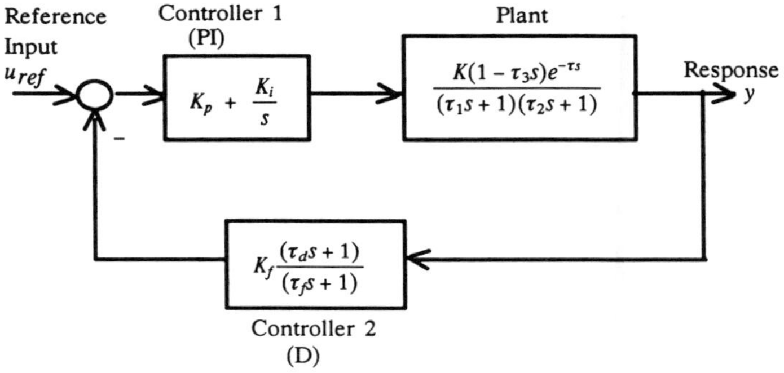

132 INTELLIGENT CONTROL: FUZZY LOGIC APPLICATIONS

Figure 7 An example of a nonminimum phase plant with transport delay and controlled by a PID servo.

time constant τ_d, is studied. The plant parameters used in this example are $K = 1.0$, $\tau_1 = 2.0$, $\tau_2 = 0.5$, $\tau_3 = 0.1$, and $\tau = 0.2$. Also, the controller parameters K_f and τ_f are fixed at 1.0 and 0.02, respectively.

Membership Function of the Rule Base

The attributes of the system response which are used in tuning the PID parameters are steady divergence, overshoot, oscillation amplitude, speed of response, and steady-state error, for a step input command. The first three attributes are primarily stability indicators and the fourth is a system bandwidth (speed or maximum frequency of interest of the response) indicator. In the present example overshoot and oscillation amplitude are treated as the same attribute, and only the latter is included in the tuning scheme. Furthermore, rise time (time to reach the desired response level for the first time) is considered as the speed of response attribute. The tuning criteria used are presented in Table 1. In a conventional scheme fuzzy terms such as "small", "moderate", "large", and "very large" must be introduced in these criteria. However, this fuzzy resolution is implicitly incorporated into the resolution relationship in the present scheme.

The membership functions of both condition and action variables are given by the triangular function

$$\mu_{X(i)}(x) = max\left(0, \frac{x - \overline{x}_{i-1}}{\overline{x}_i - \overline{x}_{i-1}}\right) \quad \text{for } x \leq \overline{x}_i$$

$$= max\left(0, \frac{\overline{x}_{i+1} - x}{\overline{x}_{i+1} - \overline{x}_i}\right) \quad \text{for } x \geq \overline{x}_i \quad (38)$$

TABLE 1
THE TUNING CRITERIA FOR THE PID SERVO

Performance attribute	Tuning action for		
	τ_d	K_p	K_i
Steady divergence	Increase	Decrease	Decrease
Overshoot/oscillations	Increase	Decrease	Decrease
Speed of response	No change	Increase	Decrease
Steady-state error	No change	Decrease	Increase

in which X stands for either E or C and $\bar{x}_i$ denotes the modal value. Note that E denotes any one of the four performance attributes that are listed in the first column of Table 1 and C denotes any one of the three tuning parameters (τ_d, K_p, and K_i) that are listed in the top row of Table 1. As discussed before, only one performance attribute and one tuning parameter are considered at a given instant of tuning and all such applicable pairs are considered sequentially. In each tuning cycle the order of the sequence is arranged according to some desirable criterion of priority. In the present example tuning actions for a given performance attribute are carried out in the order $\tau_d \rightarrow K_p \rightarrow K_i$ as dictated by the reaction time of the corresponding control actions (derivative action being faster than the integral action). Furthermore, the order in which the performance attributes are considered during tuning is arranged in the order given in Table 1 (i.e., divergence $\rightarrow$ oscillations $\rightarrow$ speed $\rightarrow$ steady-state error). This order is also justified according to some practical criterion; for example, undesirable instabilities must be corrected before adjusting the speed of response or before compensating for any steady state error. It should be mentioned that every performance attribute may not be active in all tuning cycles. For example, the two tuning actions K_p and K_i corresponding to the steady-state error attribute will not be active unless the steady state is reached, and then too unless the specified performance requirement on steady-state error is violated.

For simplicity, the resolution relationships are represented as linear functions of the index i (i.e., $m_e = 1 = m_c$ in Equations (13) and (14)). Now, in order to establish a membership function for each tuning criterion, as given by Equation (9), first we form

$$\mu_{R_i}(e,c) = min[\mu_{E(i)}(e), \mu_{C(i)}(c)]$$

which becomes

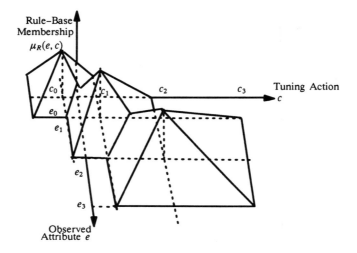

Figure 8 Membership function of a tuning criterion.

$$\mu_{R_i}(e,c) = max\left(0, min\left[\frac{e-\bar{e}_{i-1}}{\bar{e}_i - \bar{e}_{i-1}}, \frac{c-\bar{c}_{i-1}}{\bar{c}_i - \bar{c}_{i-1}}\right]\right) \quad \text{for } \begin{array}{c} e \leq \bar{e}_i \\ c \leq \bar{c}_i \end{array}$$

$$= max\left(0, min\left[\frac{e-\bar{e}_{i-1}}{\bar{e}_i - \bar{e}_{i-1}}, \frac{\bar{c}_{i+1}-c}{\bar{c}_{i+1} - \bar{c}_i}\right]\right) \quad \text{for } \begin{array}{c} e \leq \bar{e}_i \\ c > \bar{c}_i \end{array}$$

$$= max\left(0, min\left[\frac{\bar{e}_{i+1}-e}{\bar{e}_{i+1} - \bar{e}_i}, \frac{c-\bar{c}_{i-1}}{\bar{c}_i - \bar{c}_{i-1}}\right]\right) \quad \text{for } \begin{array}{c} e > \bar{e}_i \\ c \leq \bar{c}_i \end{array}$$

$$= max\left(0, min\left[\frac{\bar{e}_{i+1}-e}{\bar{e}_{i+1} - \bar{e}_i}, \frac{\bar{c}_{i+1}-c}{\bar{c}_{i+1} - \bar{c}_i}\right]\right) \quad \text{for } \begin{array}{c} e > \bar{e}_i \\ c > \bar{c}_i \end{array} \quad (39)$$

Subsequently, the sup operation is carried out over the entire range of the index i, which results in membership function of the form shown in Figure 8. This function is programmed into the tuning scheme. Because the resolution curves are linear in i, with zero offset, the modal values $\bar{e}_i$ and $\bar{c}_i$ can be expressed as $e_i = i\bar{e}$ and $c_i = i\bar{c}$, in which $\bar{e}$ and $\bar{c}$ are appropriate scaling parameters for the corresponding condition and action variables. Consequently, just one generic membership function need be programmed as a subroutine in a nondimensional form for all combinations of tuning criteria. Finally, incremental tuning actions are computed using Equation (7), via simple trapezoidal integration.

To illustrate the main steps of the tuning program further, consider the condition-action pair (divergence, τ_d). First, the scaling parameters $\bar{e}_i$ and $\bar{c}_i$ for this pair must be chosen appropriately. This is usually done through past experience: First a performance specification, $\bar{e}_{max}$, is specified for steady divergence. Then, if only a mild tuning action is desired when error exceeds $\bar{e}_{max}$ we may choose $\bar{e} \geq \bar{e}_{max}$. If stronger action is desired $\bar{e}$ may be chosen smaller than $\bar{e}_{max}$. Subsequently, the values of $\bar{e}_{max}$ and $\bar{e}$ may be adjusted in an appropriate manner through past experience (by simulation or experimentation). Here, we have used $\bar{e}_{max} = 1.0$ and $e = 0.5$ for the divergence attributes. Similarly, for τ_d a starting value τ_{d0} and a modal scaling value $\bar{c}$ must be chosen. For example, τ_{d0} may be chosen as the lowest value that will not instantaneously damage the system. Because $\bar{c}$ scales the tuning increment it should depend on both the required magnitude of tuning and the frequency at which tuning is done (tuning bandwidth). The tuning and control bandwidths are discussed in the subsection. We used $\tau_{d0} = 0.0$ and $\bar{c} = 0.005$ for the derivative action in the context of steady divergence. Once these numerical values are decided, the program steps for tuning in relation to the present condition-action pair are as follows:

1. Compute (measure) the error response e.
2. If $e/\bar{e}_{max} < 1.0$, Then No Action, and proceed to check the next performance attribute (i.e., oscillations). If $e/\bar{e}_{max} \geq 1.0$ Then Go to step 3.
3. Compute $e/\bar{e}$.
4. Call the subroutine for nondimensional membership function $\mu_R(e, c)$ of the rule base (see Figure 7), and obtain the nondimensional action-value function corresponding to the nondimensional e computed in step 3.
5. By numerical integration (according to Equation (7)) compute the centroid of the action value function obtained in step 7 (see Figure 3).
6. Scale the c value computed in step 5 by $\bar{c}$.
7. Increment τ_d by the value computed in step 6. If the hard limits of τ_d are exceeded, τ_d is set back to the old value.
8. Tune the next parameter (i.e., K_p).

Note that all computations are done on discrete (not continuous) data. Specifically, 100 data points were used for each membership function corresponding to each index value. Also, the response integration and control signal computation are done in periods of 0.02 s. This corresponds to a "control bandwidth" of 50.0 Hz.

Stability Region

For a nominal system (plant and the PID servo loop) given in Figure 7 with $K_p = 5.0$, $K_i = 2.0$, and $\tau_d = 0.01$ the stability regions were determined, assuming the first order Padé approximation for the transport delay. The stability regions were computed to be K_p: [0.445, 8.003], K_i: [0, 4.696] and τ_d: [0, 1.089]. The eigenvalues of the nominal system are $-0.2712 \pm j\, 2.0402$, -0.3875, -12.3018, and -49.5182. It follows that the nominal system is stable. Note that the first three poles are contributed by the two time constants of the plant and by the integral control action. The fourth pole results primarily from the Padé approximation to the transport delay, and the fifth pole is due mainly to the time constant τ_f in Controller 2. The tuner is expected to push the system into the stability region computed here, even when the tuning process is started outside the region (unstable system). This is illustrated in the next section.

Tuning Results

The tuner was activated with a unit step input to the system and with the initial PID parameters, $K_p = 12.0$, $K_i = 2.0$, and $\tau_d = 0.01$. Note that according to the results in the previous section, the initial system is unstable. The control bandwidth was 50 Hz and the tuning bandwidth used was 1 Hz. This means that the control signal is updated at every 0.02 s and each tuning decision is distributed over a period of 1 s. Note that even though the performance attributes are checked at the rate of the tuning bandwidth (every 1.0 s) this does not mean that tuning actions are made for each one of these attributes at this rate. For example, the actions corresponding to the steady-state error are activated if and only if the following two conditions are satisfied: (1) the steady-state is reached; (2) the steady-state error exceeds the specified tolerance value. The total duration of tuning will depend on many factors such as the plant characteristics, controller characteristics, the magnitudes of various tuning actions, and the frequency of the tuning actions. The tuning simulation was carried out for a duration of 20 s.

The response of the system with the tuner active is shown in Figure 9. Observe that the tuned condition is reached in less than 14 s in this case. It is clear that the tuner detects the stability quite early in the response and takes corrective measures. The tuner subsequently goes on to making finer adjustments, including one for steady-state error (toward the middle and end of the simulation when the steady state is reached). The control signal that goes into the plant is shown in Figure 9(e).

Figures 10(a) and (b) show the response of the system to the pulse sequence of Figure 10(c). When the PID parameters were fixed at the untuned initial values, as expected, the response became quite unstable just a fraction of the way into the first pulse, as shown in Figure 10(a). However, when tuned PID parameters were employed, the response was found to be quite satisfactory, as given in Figure

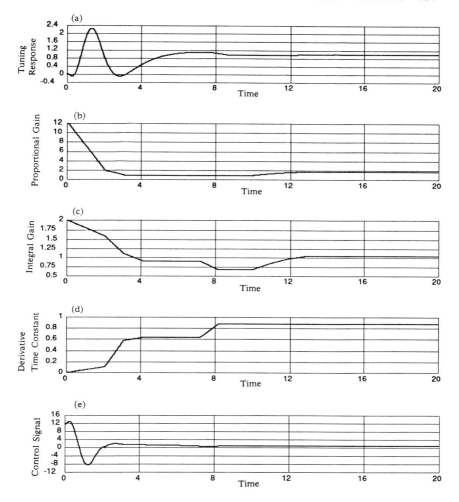

Figure 9 The tuning response of the system with a step input. (a) System response, (b) proportional gain, (c) integral gain, (d) derivative time constant, (e) control signal to the plant.

10(b). The control signal into the plant corresponding to the reference pulse command for the tuned system is shown in Figure 10(d).

SUMMARY

A knowledge-based control structure was considered in which the knowledge-based functions were relegated to a level above the direct control level. Fuzzy logic was used for knowledge representation and for inference, and the purpose of the knowledge-based system

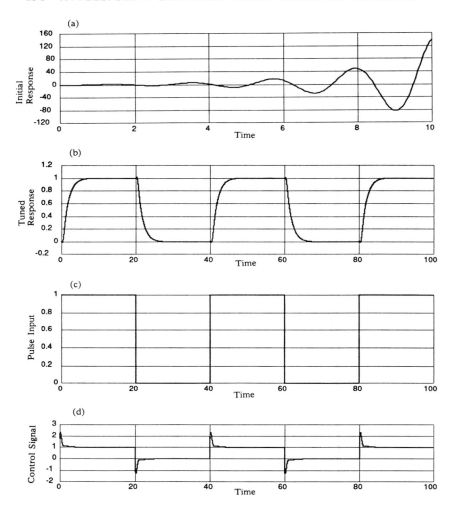

Figure 10 System response to a pulse input (a) before tuning, (b) after tuning, (c) input pulse signal, (d) control signal.

was to tune the parameters of a servo controller which occupied the lower level of the system. Fuzzy logic was not employed in direct control. This is quite realistic because often, operating experience is gained by a control engineer, not by manually sequencing a control signal and observing the response, but rather by manually tuning the controller parameters in response to the system behavior.

First the tuner was expressed using a conventional fuzzy logic formulation for both the rule base and the compositional rule of inference. Next, two concepts, *rule dissociation* and *fuzzy resolution,* were introduced. An analytical framework was built using these concepts, and in particular, by incorporating *resolution relationships*. In this manner it was possible to represent the knowledge base by a membership function in a convenient, analytic form and to vary the fuzzy resolutions of the condition and action variables without having to reformulate the problem. These are definite advantages in analyzing and implementing the knowledge-based tuner. Also, by employing nonlinear resolution relationships it would be possible to incorporate different levels of tuning weight in different regions within the acceptable range of a variable for both condition and action variables.

Stability of the overall control system was discussed in the context of the control structure considered in this chapter. In particular, sufficient conditions for analyzing stability of the overall system in the state space of the original system in the absence of the tuner were discussed. Because tuning bandwidth is typically an order of magnitude smaller than the control bandwidth, this sufficient condition should hold at least approximately in many cases of practical importance.

The total number of *min* and *max* operations that would be needed in rule base computations (i.e., to compute the membership function of the rule base) and inference computations (i.e., to apply the compositional rule of inference) was analyzed separately for both the conventional fuzzy logic formulation and the dissociated formulation of the tuner. It was found that the computational requirement increased exponentially with the total number of performance attributes and tuning parameters of the system in the conventional formulation, whereas it increased only linearly in the dissociated formulation. Consequently, substantial savings of the real-time computing effort would be possible with the dissociated formulation within the present analytical framework.

The effect of the fuzzy resolutions of a control/tuning knowledge base on the performance of a control system has been studied both analytically and experimentally. Generally speaking, the finer the resolution, the higher the response accuracy, or alternatively, the faster the response settling time. However, a higher resolution usually means a heavier computational load, and consequently smaller operating bandwidth. Moreover, beyond a certain level, increasing the fuzzy resolution may not significantly improve the system performance. Therefore, in practical applications, the fuzzy resolution must be chosen to satisfy such contradictory requirements.

A fuzzy tuning system was developed by incorporating a *resolution-switching* mechanism and was implemented in a retrofitted industrial robot. The experiments show that system responses can be improved by the resolution switching mechanism even at a low level of fuzzy resolution.

An example was given to illustrate the application of the various concepts discussed within the analytical framework of this chapter. A second order nonminimum phase plant with transport delay was considered with a PID controller, and conventional tuning criteria based on step response for such a controller were used as the knowledge base. Linear resolution relationships were employed to simplify the computer simulation. First, stability regions were computed for the nominal system (plant and a PID control loop) using the first order approximation for the transport delay. Next the controller was tuned using the fuzzy-logic tuner starting from a set of unstable values for the PID parameters. The tuning bandwidth used was 50 times smaller than the control bandwidth. The tuner was able to quickly stabilize the step response and then to make the plant behavior more accurate. The superior behavior of the tuned system was illustrated further by simulating the response to a pulse sequence.

■ PROBLEMS

1. It is argued that the fuzzy logic approach is not appropriate for low-level servo loop control. People tend to have very high expectations of fuzzy logic control, as a universal control approach. This may also be the reason why they hasten to criticize the approach. Discuss why fuzzy logic control may not be appropriate when the control objective is

    $$e = 0 \quad \text{and} \quad \dot{e} = 0$$

    where, e is the response error.

2. List several benefits of intelligent process control. If the process task can be decomposed into a succession of basic operations and the control requirement can be represented in terms of a set of subcontrollers, depending on the needed level of intelligence, suggest a suitable system structure for control and communication purposes.

3. Stability of a system may be monitored on-line in several ways. For example, one may:
    (a) identify a linear state-space (time-domain) model using input-output data and compute its eigenvalues.
    (b) identify a transfer-function (frequency domain) model using input-output data and establish relative stability (say, using gain and phase margins).
    (c) observe the trends of the process output (response) and qualitatively establish stability.

(d) inject a known disturbance to an input signal and observe the process response and establish stability by evaluating the change in response.

Discuss which of these approaches are appropriate for use in a fuzzy-logic based self-tuning system.

4. Consider a problem of decision making through fuzzy logic. As the precision of the available data (context) increases what is its implication on:
   (a) Quality of the inference
   (b) Cost of the decision making process

5. Consider the process error e, its rate $\dot{e}$, and so on, which may be monitored for control and self-tuning purposes. In self-tuning one may first use a knowledge-based preprocessor to evaluate the error variables, say with respect to a *reference model*, and arrive at response features such as stability; say,

$$(e, \dot{e}) \to A$$

Then a tuning rule base may be used, with preprocessed features as condition variables and tuning operations as actions; say,

$$A \to T$$

Alternatively, in self-tuning control, one may use an uncoupled rule base which relates individual error variables to separate tuning actions; say,

$$e \to T_1$$
$$\dot{e} \to T_2$$

But, in low-level direct (in-loop) control that uses a rule base, such approaches as given above are normally unacceptable, and instead combined rules are used; for example,

$$(e, \dot{e}) \to c$$

Discuss the rationale for this distinction.

6. Consider the single condition–single action rule base given by

$$\bigcup_i (E_i \to C_i)$$

where i denotes the ith fuzzy state (e.g., small, medium, large) of the condition variable E (e.g., response error) and C_i denotes the corresponding fuzzy state of the action variable C (e.g., control or tuning action). For a given crisp observation (measurement) $\hat{e}$ of the process, the corresponding corrective action as inferred by a conventional fuzzy logic system may be outlined as follows:

A. Determine the membership grade μ_i of the element $\hat{e}$ in the fuzzy set E_i by reading off the value $\mu_i = \mu_{E_i}(\hat{e})$ for all i.

B. Determine the fuzzy action $\hat{C}_i$ corresponding to $\hat{e}$, as inferred by the ith rule of the rule base. This is obtained by clipping $\mu_{C_i}(c)$ at the height μ_i; thus,

$$\mu_{\hat{C}_i}(c) = min(\mu_i, \mu_{C_i}(c))$$

Note: Here the *min* operation is used for "implication". Alternatively, the "*dot*" (or *product*) operation may be used as

$$\mu_{\hat{C}_i}(c) = \mu_i \mu_{C_i}(c)$$

C. Combine the inferences of all applicable rules in the rule base. Here, the *max* operation is used for "OR" ("ELSE"), to obtain the overall inference $\hat{C}$; thus,

$$\mu_{\hat{C}}(c) = \max_i \mu_{\hat{C}_i}(c)$$

D. Defuzzify $\hat{C}$ to obtain a crisp action $\hat{c}$, for instance using the centroid method.

An alternative approach would be to first form the overall rule base R as:

$$\mu_{R_i}(e,c) = min(\mu_{E_i}(e), \mu_{C_i}(c)) \text{ for rule } i$$

$$\mu_R(e,c) = \max_i \mu_{R_i}(e,c) \text{ for the complete rule base.}$$

Then, read off the value of $\mu_R(e, c)$ corresponding to the point $e = \hat{e}$, to obtain the fuzzy inference $\hat{C}$; thus,

$$\mu_{\hat{C}}(c) = \mu_R(\hat{e}, c)$$

which may be defuzzified as before.
Show that the two results are identical.
Hint: Use the *singleton function* $s(x)$ defined as

$$s(e - \hat{e}) = 1 \quad \text{for } e = \hat{e}$$
$$= 0 \text{ elsewhere.}$$

7. Knowledge-based decision making is considered to involve cycles of perception, representation, and action. In fuzzy-logic based decision making, indicate what each of these three terms represent.

REFERENCES

Astrom, K.J., Anton, J.J., and Arzen, K.E. (1986). Expert Control, *Automatica*, Vol. 22(3), pp. 277–286.

Astrom, K.J. and Hagglund, T. (1989). An Industrial PID Controller, Proc. 1989 *IFAC Symp. Adaptive System in Control and Signal Processing*, France, pp. 293–298.

Astrom, K.J. and Wittenmark, B. (1989). Adaptive Control, Addison-Wesley, Reading, MA.

de Silva, C.W. (1989). *Control Sensors and Actuators*, Prentice-Hall, Englewood Cliffs, NJ.

de Silva, C.W. and MacFarlane, A.G.J. (1989). Knowledge-Based Control Approach for Robotic Manipulators, *Int. J. Control*, Vol. 50(1), pp. 249–273.

de Silva, C.W. (1991a). Fuzzy Information and Degree of Resolution within the Context of a Control Hierarchy, *Proc. IEEE Int. Conf. Industrial Electronics, Control, and Instrumentation*, Kobe, Japan, pp. 1590–1595.

de Silva, C.W. (1991b). An Analytical Framework for Knowledge-Based Tuning of Servo Controllers, *Int. J. Eng. Appl. Artif. Intelligence*, Vol. 4, No. 3, pp. 177–189.

de Silva, C.W. (1993). Hierarchical Preprocessing of Information in Fuzzy Logic Control Applications, *Proc. 2nd IEEE Conf. on Control Applications*, Sept. 13–16, Vancouver, Canada, Vol. 1, pp. 457–461.

Dubois, D. and Prade, H. (1980). *Fuzzy Sets and Systems*. Academic Press, Orlando, FL.

He, S.Z. (1992). Design of an On-Line Rule-Adaptive Fuzzy Control System, *Proc. 1992 IEEE Int. Conf. Fuzzy Systems*, San Diego, CA, pp. 83–91.

Kosko, B. (1992). *Neural Networks and Fuzzy Systems*, Prentice-Hall, Englewood Cliffs, NJ.

Mamdani, E.H. (1977). Application of Fuzzy Logic to Approximate Reasoning Using Linguistic Synthesis, *IEEE Trans. Comp.*, C-26(12), pp. 1182–1192.

Narendra, K.S. and Annaswamy, A.M. (1989). *Stable Adaptive Systems*, Prentice-Hall, Englewood Cliffs, NJ.

Terano, T., Asai, K., and Sugeno, M. (1992). *Fuzzy Systems Theory and Its Applications*, Academic Press, San Diego, CA.

Wang, B.H. and Vachtsevanos, G. (1990). Fuzzy Associative Memories: Identification and Control of Complex Systems, *IEEE Symp. Intelligent Control*, Philadelphia, pp. 910–915.

Wu, Q.M. and de Silva, C.W. (1995). Dynamic Switching of Fuzzy Resolution in Knowledge-Based Self-Tuning Control, submitted for publication.

Zadeh, L.A. (1965). Fuzzy Sets, *Inf. Control*, Vol. 8, pp. 338.

Zadeh, L.A. (1979). A Theory of Approximate Reasoning, *Machine Intelligence*, Hayes, J. et al. (Eds.), Vol. 9, pp. 149–194.

Ziegler, J.G. and Nichols, N.B. (1942). Optimum Settings for Automatic Controllers, *Trans. ASME*, Vol. 64, pp. 759–768.

6 KNOWLEDGE-BASED CONTROL OF ROBOTS

▬ INTRODUCTION

Productivity and product quality of an automated manufacturing process rely on the accuracy of the individual manufacturing tasks such as parts transfer, assembly, welding, and inspection. In modern manufacturing workcells many of these tasks are carried out by robotic manipulators. The performance of a robotic manipulator depends considerably on the way the manipulator is controlled, and this has a direct impact on the overall performance of the manufacturing system (de Silva, 1995; de Silva and MacFarlane, 1989b; Staugaard, 1987).

Machine intelligence and robotics are related and complementary fields of study. Computer scientists are active in the development of software which can make computers perform tasks in an apparently intelligent manner. Engineers in the robotics field have achieved satisfactory results in integrating tactile sensing, vision, voice actuation, automatic ranging, and even the sense of smell into robotic systems. In short, there has been a significant effort in making robots more intelligent and human-like. Sensing, actuation, and control are the three key components of robot intelligence. Intelligent control can significantly improve the performance of a robotic manipulator. Intelligence can be incorporated into a controller in the form of a knowledge base, typically expressed as a set of rules, and an associated inference mechanism. The structure of a knowledge-based controller cannot be chosen arbitrarily. The nature of the process, control requirements, and the need for conventional "hard" (crisp) controllers must be properly considered in developing a structure for a knowledge-based control system.

The control problem associated with a robotic manipulator is not trivial, particularly in real-time control at bandwidths on the order of 100 Hz. Frequently cited reasons for this include strong nonlinearities present in manipulators (dynamic nonlinearities such as Coriolis and centrifugal accelerations, geometric nonlinearities arising from large-excursion kinematics, and physical nonlinearities such as friction and backlash), dynamic coupling between components, unplanned influences (e.g., changes in the desired trajectory for reasons such as obstacle avoidance and unexpected changes in the payload), and unknown factors (e.g., model errors and parameter uncertainties). A robotic manipulator is a representative process to which knowledge-based control could be applied. Robotics, knowledge-based control, and fuzzy logic are central ingredients of such an approach. This chapter explores those subjects which provide the relevant background.

Because robots are high-speed processes that are highly nonlinear, dynamically coupled, and often of high order, it is not adequate only to use linear servo control, if accurate performance in high-bandwidth operations is desired. Many algorithms are available for controlling a robot. Many of these control techniques depend on a dynamic model of the robot, and are termed *model-based control*. Available control algorithms include feedforward control (Asada et al., 1983; Luh et al., 1980), resolved motion control (Whitney, 1969; Wu and Paul, 1982), nonlinear feedback control (Hemami and Camana, 1976; Tokumaru and Iwai, 1972), sliding mode control (Young, 1978; Slotine, 1985), model-referenced adaptive control (Dubowski and Des Forges, 1979), and least-squares adaptive control (de Silva, 1984; de Silva and Van Winssen, 1987). Kinematic and kinetic (dynamic) formulations are useful in modeling and control of robots (Kahn and Roth, 1971; Uicker, 1965). Furthermore, computationally efficient recursive algorithms are necessary for real-time computation of dynamic variables for control (Hollerbach, 1980; de Silva, 1986; de Silva and Van Winssen, 1987).

The control techniques mentioned above fall into the category of hard-algorithmic control. They are "crisp" in the sense that control laws are presented in the form of well-defined computational algorithms that generate control actions quantitatively. In some (in particular, high-level) tasks hard-algorithmic control does not provide the desired performance in robotic manipulators. One reason for this shortcoming is that there are many factors in a manipulator, its task, and its environment that are "qualitative and fuzzy" in nature, and cannot be expressed in precise quantitative terms. For example, there is a difference in the way a human hand grasps a light bulb and the way it grasps a rubber ball. Translation of this process accurately into a control algorithm for a robot is quite difficult, however, and may involve a set of protocols which cannot be expressed as a precise analytic algorithm. These rules will represent the "knowledge" associated with the particular task and generally will be qualitative and incomplete, and could be improved through experience and learning. Knowledge-based intelligent control is important in this context. In particular, fuzzy control in which control rules may be expressed as a set of fuzzy linguistic statements has been employed for this purpose.

In its normal mode of operation a robot may be subjected to unplanned events and unfamiliar situations, and it will be required to respond intelligently to these situations. Furthermore, robot performance could be improved if the control system were able to capture and utilize past experience and available human expertise. A knowledge-based control system would be attractive in these respects. However, as some implementations show, it is not advisable to include a knowledge-based controller directly within a servo loop to generate drive signals for robot joints (see Chapter 6). Due to the nature of the particular process (a robot), even if the processing speed were not a limitation, a human expert would rely on a crisp algorithm for this purpose of control signal generation. Because human experts are quite effective in tuning operations, constantly improving their skills through experience, it is intuitively clear that a knowledge-based controller for a robot may be more effective in a supervisory (monitoring and tuning) capacity than in a direct servo-control capacity.

Consider a control structure that has a three-level hierarchy. The lowest level will consist of a crisp controller containing a group of conventional servo controllers which are closed around a high-speed decoupling and linearizing controller. At level two there will be a knowledge-based system of the forward production-system type, which will consist of an intelligent observer (a servo expert) for each degree of freedom of the robot. The soft controller at the top layer of the control structure will function as an intelligent tuner for the "hard" controller at the lowest level. Fuzzy logic will be used for the development of the intelligent tuner. The group of intelligent preprocessors in level two will provide inferences regarding a series of attributes of the process response with respect to a set of performance specifications. This information will then be supplied to the fuzzy controller in the top level. The fuzzy controller will carry knowledge, represented as linguistic statements obtained from human experts, which may contain soft, qualitative terms for tuning the joint servos. Based on measurements, inferences from the servo experts, and possibly other types of external information, the fuzzy controller will generate tuning commands for the servos at the lowest (crisp control) level. Control structure of this type is discussed later in this chapter, with specific application to robots (de Silva and MacFarlane, 1989b).

ROBOTIC CONTROL SYSTEM

A robot is a mechanical manipulator that can be programmed to perform various physical tasks. Programmability and the associated task flexibility are necessary characteristics for a robot, according to this commonly used definition. Autonomous operation would be desirable under special circumstances, for example, in space-based or extremely hazardous activities. Furthermore, a robotic task might be complex and the environment might be unstructured to the extent that some degree of intelligence would be required for satisfactory performance of the task. There is an increasing awareness of this (Staigaard, 1987; de Silva and MacFarlane, 1989b), and the trend has been to include intelligence, which would encompass abilities to perceive, reason, learn, and infer from incomplete information, as a requirement in characterizing a robot.

A robot may be interpreted as a control system. Its basic functional components are the structural skeleton of the robot; the actuator system, which drives the robot; the sensor system, which measures signals for performance monitoring, task learning and playback, and control (both feedback and feedforward); the signal modification system for functions such as signal conversion (e.g., digital to analog and analog to digital), filtering, amplification, modulation, and demodulation; and the direct digital controller, which generates drive signals for the actuator system so as to reduce response error (de Silva, 1985, 1989). Higher level activities such as task description, procedural decomposition, sub-task allocation, path planning, activity coordination, and supervisory control are also integral within the control system.

A schematic diagram of such a control system, primarily from the hardware perspective, is given in Figure 1. In fact, in a typical scenario, the robot is merely an integral component of a workcell or even a multiple workcell production system. Functions of an overall system may be represented by a hierarchical structure, as shown in Figure 2 (de Silva, 1991a; 1995). Basically, the actions of the robot must be planned and properly coordinated depending on the tasks to be accomplished and on the environment within which the robot would operate.

Control requirements for a robot could be identified by considering the activities within each level of a system architecture, as in Figure 2. Process tasks are defined at the system-host level and decomposed into subtasks and associated procedures. This information is downloaded into the cell-host computer, which manages the workcell operation. The cell host assigns subtasks to all the components (robots, machine tools, parts transfer devices, etc.) and coordinates their operation. In particular, the cell host must provide a sequence of robot operations (end-effector positions, orientations,

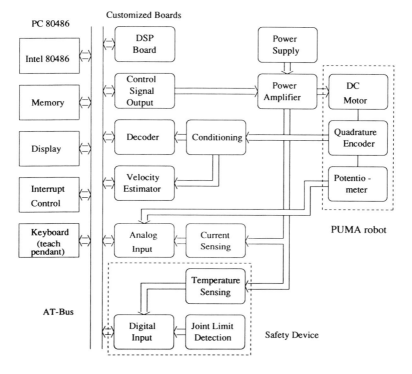

Figure 1 The control system of a research robot.

and movements), expressed in a Cartesian coordinate frame of the task space of the robot, along with the necessary timing signals for task synchronization and also for object-handling information (e.g., pick, place, tool-change). The associated trajectories must be properly planned for smooth operation. Because the robot is driven by the actuators at its joints, the task-space Cartesian information must be transformed into the joint space, which provides the reference commands for actuating the joints. The low-level controller, using proper feedback sensory signals, ensures that the robot joints are moved faithfully according to these reference commands.

■ APPLICATION TO ROBOTS

Strong nonlinearities, dynamic coupling, high-order dynamics, and unknown influences have been mentioned as reasons for control difficulties associated with robotic manipulators. Several crisp algorithms intended for handling these difficulties were outlined in Chapter 1, and their shortcomings pointed out. Knowledge-based control, in particular fuzzy control, is a potentially powerful approach

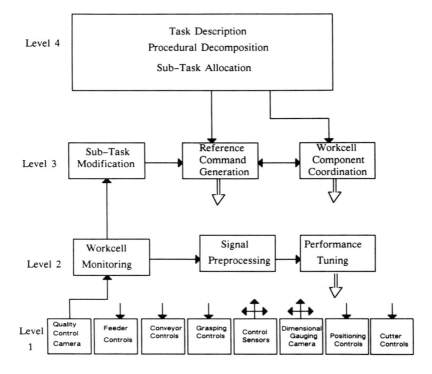

Figure 2 The hierarchical structure of a robotic system.

which can capture human experience and expertise in controlling complex processes, thereby circumventing many of the shortcomings of hard-algorithmic control.

Several conditions have been commonly identified (de Silva, 1993) under which fuzzy logic may be effectively used in controlling a plant. For example, if a plant is complex and ill-defined, analytical modeling becomes difficult. Furthermore, experimental modeling might be infeasible as well, particularly if the inputs and the outputs are not accessible for accurate sensing. Under these circumstances, model-based control approaches will be unsuitable, and control techniques that employ model identification (e.g., some classes of adaptive control and nonlinear feedback control) will fail. Then, assuming that sufficient knowledge or experience on the plant is available that could be expressed as a set of control protocols, perhaps containing fuzzy terms, one is presented with an excellent opportunity to attempt fuzzy logic control for the plant. Because the human experience in controlling a plant is typically of a low-bandwidth and supervisory nature, one may conclude that direct, in-the-loop control using fuzzy logic is not well justified unless the time constants of the plant are high (e.g., 1.0 s or more) and the required control bandwidth is sufficiently low (e.g., 1.0 Hz or less). (See Chapters 4 and 5.)

A survey of the existing applications of fuzzy logic in robotic control reveals that these applications could be grouped into the following four main categories:

1. In-loop direct control
2. Information filtering and preprocessing
3. System monitoring
4. Task design.

Most of the research effort in process control has been concentrated on crisp, analytic algorithms, while in the realm of knowledge-based control, attention has been limited primarily to an exclusive use of soft and qualitative techniques. In robot control research, in particular, a few researchers have used soft techniques at the servo level (Scharf and Mandic, 1985; Hirota et al., 1985). Specifically, conventional fuzzy control has been implemented directly in the servo loops. Results often show (Scharf and Mandic, 1985) that direct servo control outperforms sophisticated fuzzy control when implemented in this manner.

In robot control the reason should be intuitively clear why knowledge-based controllers often do not give satisfactory results. For a robot, assuming that speed of operation is no obstacle, suppose that a human expert is called upon to generate control signals for the joint actuators. It is quite likely that an expert will resort to a crisp algorithm to accomplish this because it would be virtually impossible to generate appropriate drive signals heuristically. On the other hand, if the expert interacts with the control loop in an advisory capacity, perhaps for tuning the control parameters, then generally the performance should improve. This possibility is supported by experience in the process control industry. This intuitive reasoning provides a rationale for the hierarchical control structure for robot control (de Silva and MacFarlane, 1989b).

The scope of this chapter is the integration of crisp algorithmic control and soft knowledge-based control in the context of robotic manipulator control. A hierarchical control structure, which has a high-speed "hard" controller at the lowest level and an intelligent observer and a tuner at the upper levels, is described. The chapter explores the rationale for the specific control structure proposed, and develops the necessary qualitative and analytical foundation for the control structure. To demonstrate the development and feasibility of this control structure, an example application has been programmed on a SUN workstation. A two degree-of-freedom robot has been simulated as a separate UNIX process using a program written in C. The servo experts have been developed using the MUSE AI toolkit. The top-level fuzzy controller has been developed separately using a valid set of linguistic tuning rules for PID servos, and implemented as a set of decision tables within the knowledge-based controller. The robot simulator has been interfaced to the servo experts using the UNIX socket facility along with the "Channel" objects of MUSE and interface programs written in PopTalk. The fuzzy controller has been interfaced with the robot simulator using the "Stream" objects of MUSE. The control structure has been evaluated on the basis of the simulation results.

▰ IN-LOOP DIRECT CONTROL

In the context of a hierarchical control system of the type shown in Figure 2, it is reasonable to presume that the logical use of fuzzy logic would be for high-level control tasks. This argument may be justified in many ways; for example, higher level tasks are relatively complex, ill-defined, would not be supported by precise sensory signals, and would require a relatively higher degree of intelligence in the associated decision making process. This aspect is further studied in Chapter 8. Even though some effort has gone into incorporating knowledge-based control at the system level, (e.g., in a flexible manufacturing cell), much of the available work is devoted to adding a soft controller directly into the servo loop of a robot. This direct in-loop application of fuzzy logic control is addressed next (de Silva, 1995).

A common application of fuzzy logic control in robots is within the joint-servo loops as an integral part of the direct-digital controller (Mamdani, 1977; Procyk and Mamdani, 1979; Tong, 1977). It has taken several different forms, but a common architecture is shown in Figure 3. Here, the fuzzy logic controller "infers" the control action (U) for a joint actuator of the robot on the basis of the observed (measured) values of the position error (E) and the change in error (CE). Typically, a triangular or trapezoidal set of membership functions is used for these variables, as given in Table 1 (Part a). A fixed set of rules may be employed for decision making. Alternatively, a self-organizing rule base may be utilized, as presented in Table 1. In this latter case one starts with a preliminary rule base, a performance enhancement table (Table 1, Part b), and an estimate for the delay time of the system (determined through experience, experimentation, etc.), which is expressed as the number of sampling periods (m). The main steps involved in determining the control action are

1. Read from the table, the enhancement V corresponding to the current context variables (E, CE).
2. Consider the control rule R' that was applied m samples ago, with the corresponding control action U.

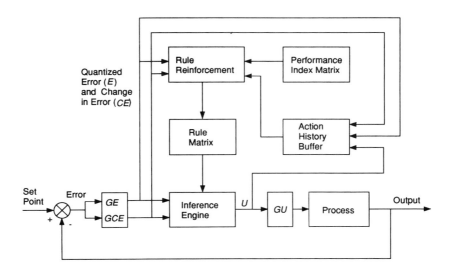

Figure 3 Application of fuzzy logic control in a servo loop of a robot (process).

3. Enhance the control rule R' to a new rule R'' with a modified control action of $U + V$.
4. Modify the existing rule base R according to:

$$R_{new} = [R \text{ And } (\text{Not } R')] \text{ Or } R'' \tag{1}$$

5. Determine the present control action using R_{new}.
6. For the next control cycle (sample period) repeat Steps 1 through 5, until the end is reached.

Note that the rule base evolves (self-organizes) during the control process. For example, an initial rule base with zero entries (null control actions) developed into that shown in Table 1 (Part c), in the tenth run of a simulation. The scaling factors and gains (GE, CGE, GU) must be properly chosen or adjusted for improved performance. There is some flexibility in choosing the membership functions (Table 1, Part a) as well.

An early application of fuzzy logic for direct control or robots was discussed by Hirota et al. (1985). A self-organizing fuzzy controller has been successfully applied to an experimental robot by Scharf and Mandic (1985). Direct fuzzy control has been applied to a simulation of a walking robot and the results have been favorably compared with those using an adaptive controller by De Yong et al.

TABLE 1
PARAMETERS OF A SELF-ORGANIZING FUZZY LOGIC CONTROLLER FOR A ROBOT

(a) Membership Functions for Error (E), Change in Error (CE), and Control Action (U)

Fuzzy value	Membership function												
	−6	−5	−4	−3	−2	−1	0	1	2	3	4	5	6
+ Very large	0	0	0	0	0	0	0	0	0	0	0.3	0.7	1
+ Large	0	0	0	0	0	0	0	0	0	0.3	0.7	1	0.7
+ Somewhat large	0	0	0	0	0	0	0	0	0.3	0.7	1	0.7	0.3
+ Medium	0	0	0	0	0	0	0	0.3	0.7	1	0.7	0.3	0
+ Small	0	0	0	0	0	0	0.3	0.7	1	0.7	0.3	0	0
+ Very small	0	0	0	0	0	0.3	0.7	1	0.7	0.3	0	0	0
Zero	0	0	0	0	0.3	0.7	1	0.7	0.3	0	0	0	0
− Very small	0	0	0	0.3	0.7	1	0.7	0.3	0	0	0	0	0
− Small	0	0	0.3	0.7	1	0.7	0.3	0	0	0	0	0	0
− Medium	0	0.3	0.7	1	0.7	0.3	0	0	0	0	0	0	0
− Somewhat large	0.3	0.7	1	0.7	0.3	0	0	0	0	0	0	0	0
− Large	0.7	1	0.7	0.3	0	0	0	0	0	0	0	0	0
− Very large	1	0.7	0.3	0	0	0	0	0	0	0	0	0	0

(b) Performance Enhancement Table

							CE						
		−6	−5	−4	−3	−2	−1	0	1	2	3	4	5	6
	−6	0	0	−1	−2	−3	−4	−6	−6	−6	−6	−6	−6	−6
	−5	0	0	0	−1	−2	−3	−4	−4	−5	−5	−6	−6	−6
	−4	0	0	0	0	−1	−2	−3	−3	−4	−5	−5	−6	−6
	−3	+1	0	0	0	0	−1	−2	−2	−3	−4	−5	−5	−6
	−2	+2	+1	0	0	0	0	−1	−1	−2	−3	−4	−5	−6
	−1	+3	+2	+1	0	0	0	−1	−1	−1	−2	−3	−4	−5
E	0	+4	+3	+2	+1	+1	0	0	0	−1	−1	−2	−3	−4
	+1	+5	+4	+3	+2	+1	+1	+1	0	0	0	−1	−2	−3
	+2	+6	+5	+4	+3	+2	+1	+1	0	0	0	0	−1	−2
	+3	+6	+5	+5	+4	+3	+2	+2	+1	0	0	0	0	−1
	+4	+6	+6	+5	+5	+4	+4	+3	+2	+1	0	0	0	0
	+5	+6	+6	+6	+5	+5	+4	+4	+3	+2	+1	0	0	0
	+6	+6	+6	+6	+6	+6	+6	+6	+4	+3	+2	+1	0	0

(c) Rule Base in the 10th Run

							CE						
		−6	−5	−4	−3	−2	−1	0	1	2	3	4	5	6
	−6	0	0	−3	−4	−4	−6	−6	−6	−6	−6	−6	−6	−6
	−5	0	0	0	0	0	−4	−5	−6	−6	−6	0	0	−6
	−4	0	0	0	0	−1	0	−3	−5	−6	−6	−5	0	−6
	−3	0	2	0	0	0	−2	−5	−4	−6	0	−6	−6	−6
	−2	0	3	0	0	0	0	0	−4	0	0	0	−6
	−1	0	4	3	0	1	0	−2	−4	−2	0	0	0	−6
E	0	0	0	0	0	3	0	2	0	0	−1	0	0	−6
	1	0	5	0	0	2	4	2	0	0	−1	0	0	0
	2	0	5	0	0	0	0	0	0	0	0	0	0	−6
	3	0	0	0	0	0	0	0	0	0	0	0	0	0
	4	0	0	6	0	0	0	0	0	0	0	0	0	−4
	5	0	0	6	0	0	0	0	0	0	0	0	0	−6
	6	6	0	0	6	6	6	6	6	5	4	3	0	6

(1992). Watanabe et al. (1993) applied a direct fuzzy controller in the trajectory control of a mobile robot. Some parameters of the rules were "tuned" using a neural network, and simulation results were given. Direct fuzzy control has been applied to a two-link rigid robot by Lim and Hiyama (1991). In this work a proportional plus integral (PPI) controller was employed for fast response and zero steady-state error, and the fuzzy logic controller was used in the conventional sense as a proportional plus derivative (PPD) controller to achieve adequate damping and satisfactory transient response. Lin and Lee (1993) applied direct fuzzy control to a single-link flexible robot. In addition, the nonlinear effects were compensated for using a "nonlinear effects negotiator", and some simulation results were given. An experimental investigation of a laboratory robot under direct fuzzy control was given by Kumbla and Jamshidi (1993). The performance was compared to that which used conventional PPD control, and was found to be superior. The two joints of a robot were directly controlled using fuzzy logic by Benerjee and Woo (1993). Simulations were developed in a specific software environment, and good performance was demonstrated.

One obvious drawback in the robotic applications outlined above is that the fuzzy control laws are implemented at the lowest layer: within a servo or direct digital control (DDC) loop, generating control signals for process actuation directly through fuzzy inference. In high-bandwidth processes such as robotic manipulators this form of fuzzy control implementation would require very fast and accurate control in the presence of strong nonlinearities and dynamic coupling. Another drawback of this direct implementation of fuzzy logic control is that the control signals are derived from inferences which are fuzzy, thereby directly introducing low-resolution errors into the control signals. A third argument against the conventional, low-level implementation of fuzzy control is that in a high-speed process, human experience is gained not through manual, on-line generation of control signals in response to process outputs, but typically through performing parameter adjustments and tuning (manual, adaptive control) operations. Hence it can be argued that in this case the protocols for generating control signals are established not through direct experience, but by some form of correlation. Of course, it is possible to manually examine input-output data recorded from a high-speed process, but on-line human interaction is not involved, and again the experience gained would be indirect and somewhat artificial. It is rather unfortunate, then, that a majority of the fuzzy logic applications in low-level direct control of robots are limited to computer simulations. Furthermore, the cases considered are situations in which dynamic modeling itself is not difficult, thereby defeating at least one purpose of choosing fuzzy logic.

Other applications are found in which fuzzy logic has been employed to "fuzzify", at least in part, a low-level conventional control algorithm. Examples include fuzzy, model-referenced adaptive control (MRAC) (Behmenburg, 1993), in which the conventional MRAC scheme (Dubowski and Des Forges, 1979) is employed to adapt an in-loop fuzzy controller; and fuzzy, sliding-mode control (SMC), in which a fuzzy rule base is employed to switch between the sliding surfaces in a conventional SMC scheme (Young, 1978). These techniques have been applied to robotic-type actuators and may be easily extended to complete robots.

HIGH-LEVEL FUZZY CONTROL

Knowledge and artificial intelligence may be incorporated at all levels of a robot-integrated control system of the type shown in Figure 2. Even though the difficulties due to friction and backlash can be significant, dynamic modeling of a robot alone is not a serious problem. Also, conventional proportional-integral-derivative (PID) control has produced satisfactory performance in robots operating in structured environments. Model-based control techniques have further improved this performance. It is the nature of the task and the environment within which a robot must operate that will call for a high degree of intelligence with necessary knowledge for controlling the robotic activity. For example, to carry out the command, "Move the end effector from point A to point B along a straight-line in 2.0 s", a robot should not need intelligent control, assuming that the workspace is free (no obstacles) and the actuator capacity of the robot is not exceeded. However, if the command is to "Assemble a pump", then it would need sophisticated control, particularly if the parts are not properly organized and some parts might be either missing or would not match. Fuzzy logic, as a robot control method, is more appropriate in the upper levels of the hierarchical system of Figure 2. Several applications of this type are considered in this section.

Information Preprocessing

Learning may be interpreted as repeated preprocessing of information through an intelligent sensor. Consider the model shown in Figure 4 for a knowledge-based hierarchical control system (de Silva, 1991a). This system will be studied in more detail in Chapter 8. Level n of the hierarchy satisfies the relation:

$$\mathbf{d}_n = \mathbf{G}_n[(\mathbf{F}_{n,n} \otimes \mathbf{i}_n) \oplus (\mathbf{F}_{n,n-1} \otimes \mathbf{i}_{n-1}) \oplus \cdots \oplus (\mathbf{F}_{n,1} \otimes \mathbf{i}_1) \oplus \mathbf{d}_{n+1}] \quad (2)$$

where:
$\mathbf{i}_k$ = sensory information (feedback) from level k

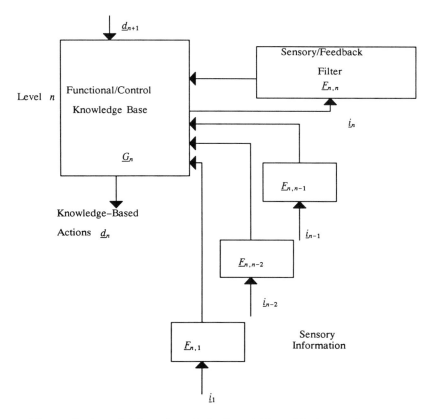

Figure 4 A model for a knowledge-based hierarchical control system.

- $\mathbf{d}_j$ = inference (control actions including planning/reference commands) at level j
- $\mathbf{G}_n$ = knowledge system in level n
- $\mathbf{F}_{n,i}$ = preprocessor for information from level i to be used by $\mathbf{G}_n$
- $\otimes$ = a generalized transitional operator
- $\oplus$ = a generalized combinational operator

Information generated at low levels generally have a high resolution and tend to be less fuzzy. Such information may not be needed in its entirety at high levels, and also may be incompatible with the knowledge systems that deal with such information at these upper layers. For example, the encoder signals from the joint motors of a robot are essential for servo control (low-level direct control), but for system tuning purposes, the complete signal would not be needed. It must be preprocessed to obtain, e.g., peak values, offset, speed, and the significant frequency bands. In addition to purely quantitative (analytical) processing, some qualitative preprocessing (for example, through past experience and heuristics) might be useful. In such a case a low-level sensor and a knowledge-based preprocessor, as modeled in Figure 4, may be termed an "intelligent sensor". This model is explored further by de Silva (1993).

We now consider some applications of fuzzy logic for information preprocessing or filtering in robotic control. A hierarchical system for robot control is given in de Silva and MacFarlane (1988, 1989a). In the associated control structure there is a knowledge system for each degree of freedom of the manipulator. This knowledge system is termed a *servo expert*, which monitors the error response of the particular degree of freedom of the manipulator. These monitored data form the context for the servo expert. Inferences are made according to the rules in the knowledge base independently, without considering the response of the remaining joints. Typically, the error response is interpreted as a response from an independent second order process. Expertise for interpreting such responses should not be difficult to gain, particularly from process control practice. For example, a desirable response can be specified in terms of appropriate measures for characteristics such as error, speed of response or bandwidth, and stability (de Silva, 1991b). It is clear that the servo experts serve as intelligent sensors/preprocessors in this application, and their knowledge base may be represented as a set of fuzzy rules.

When two or more mobile robots work cooperatively, tracking of one robot by another may be treated as an intelligent preprocessing task. In the work reported in Ji et al. (1993), ultrasonic sensors are used by robots to track or avoid other robots. The ultrasonic signals are preprocessed to obtain the necessary control commands for navigation.

In robotic assembly it is useful to identify and correctly locate the necessary parts to be assembled. Fuzzy logic is employed in Pham and Hafeez (1992) to preprocess the information from a mechanical vibration sensor for object location. The sensor is mounted at the end-effector of a robot, and consists of a rigid platform attached to a flexible column. When an object is held rigidly on the platform, the associated natural frequency changes, and this represents the sensory information. There are inherent problems of indeterminacy and noise. Preprocessing using a set of fuzzy rules has produced good results, as shown through experimentation.

System Monitoring

System monitoring is intimately related to information preprocessing because the latter process is usually a prerequisite for the former. Monitoring of a robotic system may involve performance evaluation of a single robot or a group of components in an entire workcell. In both situations similar system architecture and procedures may be used. The purpose of system monitoring at a high level is to determine whether the robotic system performs according to the plan, and if not, establish the course of action to improve the situation. This is different than sensing for low-level control, particularly in view of the high-level (intelligent) decision making that is involved in high-level control, which would use a knowledge system and preprocessed information from lower levels of the system.

A three-level hierarchical system for monitoring and self-tuning of a robot is described in (de Silva and MacFarlane (1989a) and that of a complete workcell is presented by de Silva and co-workers (1992; de Silva and Wickramarachchi, 1992). Information from the sensors of the workcell components are preprocessed so as to be compatible with the knowledge system of system monitoring. The knowledge base is expressed as a set of fuzzy rules that relate the process conditions to a set of tuning actions or system adjustments, with the objective of maintaining a high level of performance. The approach has been successfully applied to a laboratory workcell for fish processing.

A three-level hierarchical system, similar in concept to the above system, is presented by Cotsaftis (1993) for robotic systems. As in the previous structure, the lowest layer is dedicated to conventional, direct control. The intermediate layer carries out learning (intelligent sensing) functions. The top layer has a qualitative reasoning mechanism based on fuzzy logic, and performs decisional control.

Task Design
A robotic task may be decomposed into a set of procedures. This decomposition corresponds to task design and occupies the uppermost layer of the hierarchical system shown in Figure 2. The procedures themselves may be arranged in a hierarchy, with those at the lowest layer needing the least amount of intelligent decision making/control. Task decomposition, component allocation, trajectory planning, and interaction with an unstructured environment may be treated within the category of task design.

Some work has been carried out in applying fuzzy logic for this top level of robot activity. Inoue et al. (1993) discussed an approach based on fuzzy-logic reasoning for selecting an appropriate type of robotic manipulator for a given task. This approach may be useful for assigning a task to a robot, in controlling a self-reconfigurable robot, and also in robotic design. Examples are presented in the area of robotic spray painting.

Motion planning or trajectory planning of robots is another area within the present category in which fuzzy logic has been applied. Bagchi and Hatwal (1992) present a fuzzy logic-based technique for motion planning of a planar robot with redundant kinematics. Obstacle avoidance and task optimization (de Silva et al., 1988) are possible when a robot has redundant degrees of freedom. Fuzzy logic is used to reason the presence of an obstacle (including link overlap), using sensory information (preprocessing). The resulting context is used to take appropriate actions in following the desired trajectory. Moving obstacles are considered in this work, and some simulation results are presented. Path planning is done on line in this case.

Solution of the inverse-kinematics problem is useful in motion planning and in controlling a robot. Robotic tasks are usually specified as end-effector motions with respect to a Cartesian coordinate frame in the workspace. However, the robot is actuated at its joints by applying direct control actions. It follows that for control purposes, end-effector trajectories must be transformed into joint trajectories, which are expressed in the joint space of the robot. This is the inverse-kinematics problem. A closed form and unique solution exists only for a limited class of robot, and the solution can be complex and time consuming in general. For redundant robots, for example, an infinite number of solutions will exist, and some form of optimization or pseudo-inversion would be needed. Humans, however, use their arms to move, position, and manipulate objects very accurately and quickly, without using accurate relationships of inverse kinematics. They use experience, heuristics, and successive correction through visual/sensory feedback for this purpose. The inverse-kinematics problem is, then, quite amenable to a fuzzy-logic based solution. This concept is used in Xu and Hechyba (1993). A more analytical approach, using the extension principle (Zadeh, 1975; Dubois and Prade, 1980) has been presented for this problem by Kim and Lee (1993).

Another technique based on fuzzy logic for robotic motion control in the presence of obstacles is given by Zhou and Raju (1993). Robots with redundant kinematics are considered. A hierarchical structure is used to reduce the number of fuzzy rules needed for control. In the scheme the parameters of influence are arranged in a hierarchical manner, and the control actions for executing a particular movement are carried out starting with the lowest level and systematically proceeding to the top. Computer simulations for a planar redundant robot are presented to illustrate the approach.

■ CONTROL HIERARCHY

The control structure described now is an attempt to combine the advantages of crisp algorithmic control and knowledge-based soft control for the control of robotic manipulators. The approach was developed by de Silva and MacFarlane (1989a, b). In particular, crisp algorithms are restricted to high-bandwidth control loops at the servo

KNOWLEDGE-BASED CONTROL OF ROBOTS 161

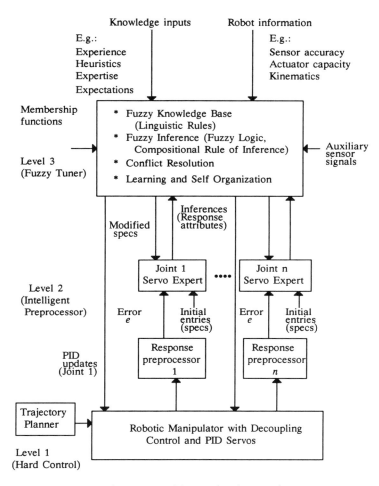

Figure 5 The proposed hierarchical control structure.

level, and knowledge-based control is introduced at higher levels in a control hierarchy in order to effectively monitor the robot performance and to tune the conventional crisp controllers at the servo level. Such a structure is expected to be superior to the use of soft control in servo-level control loops, as has been done in earlier applications of fuzzy control to robotic manipulators.

A schematic representation of the control structure is shown in Figure 5. It consists of three control levels. The lowest level corresponds to the servo controllers at the robot joints (degrees of freedom). The joint servo loops are closed around a nonlinear feedback controller which decouples and linearizes the robotic manipulator. The robot trajectory is specified by the task. This desired trajectory is generated by the trajectory planner and resolved into desired joint trajectories, and these joint trajectories then form the reference inputs to the joint servos. The joint servos are assumed to be of the PID type. They are tuned in coordination by a knowledge-based fuzzy controller at the third (highest) level of the control structure. The second layer consists of a set of servo experts. These knowledge-based controllers are used to distribute the soft control activity among the degrees of freedom of the manipulator. There is one servo expert for each joint servo. It monitors the response of the particular degree of freedom, makes inferences as to the performance of that joint, with respect to a set of control specifications, and passes that information to the fuzzy controller, without regard to the remaining joints. The fuzzy controller uses this information, and possibly other forms of data and external knowledge, to determine appropriate modifications to the controller parameters in the servo level.

Decoupling and Servo Layer
Simple linear servo control is known to be inadequate for transient and high-speed operation of robots (Horn and Raibert, 1978; Van Brussels and Vastmans, 1984; de Silva, 1984). Nevertheless, past experience of servo control in process applications is extensive, and this is true to some degree for industrial robots because servo control is the primary means of control in the present generation of commercial robots. This knowledge can be captured by a knowledge-based controller and used in servo tuning. For this type of control to be effective, however, nonlinearities and dynamic coupling must be compensated faster than the control bandwidth at the servo level. Therefore, a linearizing and decoupling controller is implemented inside the joint servo loops. This low-level controller has a recursive algorithm (de Silva and MacFarlane, 1989b) which will enable the real-time implementation of the controller.

Servo Expert Layer
In the hierarchical control structure described in this chapter there is a knowledge system for each degree of freedom of the manipulator. This knowledge system is termed a *servo expert*. Each servo expert will be implemented as a *forward production system* (FPS), having a knowledge base containing forward production rules, several databases carrying data or context, and an inference engine for controlling the overall operation.

The servo expert monitors the error response of the particular degree of freedom of the robot. These monitored data form the context for the servo expert. Inferences are made independently according to the rules in the knowledge base without considering the response of the remaining joints. Typically, the error response is interpreted as a response from an independent second order process. Expertise for interpreting such responses should not be difficult to gain, particularly from process control practice. For example, a desirable response can be specified in terms of appropriate measures for the following characteristics:

1. Error
2. Speed of response or bandwidth
3. Stability

The fuzzy controller decides what actions should be taken, based on the inferences made by the servo experts. For example, if there appears to be a steady offset in the response, increasing the integral rate constant might be the solution; if the response appears slow, the natural frequency constant might have to be increased; an oscillatory response might be corrected by increasing the derivative time constant. Note that the specifications themselves might be updated by the upper-level fuzzy controller, depending on the operating conditions and performance requirements. If enough computing power is available, input-output data could be analyzed and rules similar to Ziegler-Nichols tuning criteria, as commonly utilized in commercial knowledge-based servo tuners (PROTUNER, 1984) could be implemented in the knowledge base of the servo expert. Alternatively, knowledge engineering procedures that are employed in the development of expert systems could be followed to capture the knowledge of a group of experts in the servo control of robotic manipulators, to develop the rules *ab initio.*

If modeling errors are negligible, and if the control hardware and software in level 1 are implemented perfectly, then individual joints should behave as independent simple oscillators with PID control. These are ideal goals, and in practice the presence of some dynamic coupling and nonlinear behavior will be unavoidable. Furthermore, any disturbances, unknown influences, and software and hardware errors and delays will degrade the performance of the controller in level 1. Hence, inferences which are made by the individual servo experts will have to be further evaluated before making adjustments to the servos. This is one of the tasks which will be accomplished at the top layer (fuzzy control level) of the hierarchy.

There are particular reasons for separating the servo expert level and the fuzzy control level. By attaching a knowledge-based controller to each degree of freedom of a robotic manipulator, we distribute intelligence among crucial locations of the robot. In this manner, localized monitoring can be done quickly and efficiently. Furthermore, priorities could be attached to the inferences which are made by the servo experts. By so doing, the fuzzy controller will be able to act quickly on troubled joints without waiting for inferences from other joints, particularly under emergency conditions. Under normal conditions, when the fuzzy controller has ample time to make its inferences, information from all servo experts at the lower layer will be properly evaluated. In any event, it is necessary to delegate the final tuning decisions for individual joints to the fuzzy controller, in which appropriate linguistic rules for servo control can be conveniently implemented. For example, the servo expert can supply to the fuzzy controller, information on several attributes of the joint response without actually determining adjustments to the PID controller. In this sense, the servo experts act as intelligent preprocessors for the fuzzy tuner.

Fuzzy Control Layer
It can be shown that by continuously adjusting (ideally at an infinite bandwidth) the servo parameters of a robotic manipulator, the joints can be made to respond similar to simple oscillators. This is the idea behind MRAC of robots (Kornblugh, 1984). The disturbance terms **d** represent spill-over nonlinearities and dynamic coupling present in the robot system with a nonlinear feedback controller (de Silva and MacFarlane, 1989b). In theory, it is possible to completely eliminate **d** by adjusting the PID parameters. This is an aim of the fuzzy controller.

The primary objective of the fuzzy controller is to use its knowledge base (a set of linguistic fuzzy rules) to update the joint servo parameters, on the basis of the inferences from the servo experts, and other information that might be available. Generally, the fuzzy tuner is able to add, delete, or modify its own rule base and the rule bases of the servo experts. In addition, parameters of the nonlinear feedback controller could be adjusted, for example, by using a measure for any dynamic coupling and nonlinearity which remains after nonlinear feedback. Furthermore, the performance specifications used by the servo experts might have to be updated during operation. It follows that the fuzzy controller performs learning and self-organization functions as well as conflict resolution functions.

Because the servo experts submit their inferences to the fuzzy controller, the decision as to whether to completely accept the recommendations that are made by a servo expert, or to modify or reject them on the basis of other information that might be available is a task of conflict resolution. For example, if external sensors are used to monitor the response of the end-effector, and that information is supplied to the fuzzy controller, an evaluation of the end-effector error can be used to determine a probable cause for that error. The fuzzy controller might be able to associate one or more joints with this error. As an example, suppose that the end-effector trajectory has an offset error in the vertical direction. From kinematic considerations alone, it might be possible to determine which degrees of freedom are effective in correcting this error. If, for instance, there is a prismatic joint that is positioned vertically at the time, then that joint might be the best choice for the corrective action.

Available knowledge about joint sensors and actuators (e.g., resolutions, capacities) can also be used in decisions of this nature.

Information that is gathered by the fuzzy controller over a sufficiently long period of time can be used to modify performance parameters such as the response specifications which are used by the servo experts. This information gathering can be considered as learning. For example, by relaxing some specifications temporarily and by observing the resulting overall performance (at the end-effector), it should be possible to determine whether the servo requirements are overspecified. The frequency of joint error monitoring and of PID tuning, can also be changed through learning. Self-organization is a result of learning. In particular, modifications made to the rules and decision tables on the basis of new information and experience can be described as *self organization*.

Inferences and actions which are made at the top level of the controller are equivalent to those which could have been made by a human expert. The associated knowledge can be expressed as a set of linguistic rules, and these form fuzzy information. The knowledge base of the fuzzy controller consists of fuzzy rules and fuzzy relations. Fuzzy information is stored as membership functions. The *extension principle* (see Chapter 3) is the primary means of extracting information from fuzzy relations. When nonfuzzy information, such as sensor readings, is available for a fuzzy quantity, the associated membership function must be represented in the variable form so that the peak could be shifted accordingly to the available nonfuzzy information, so as to form a fuzzy singleton.

Tuning decisions for PID control parameters made by a human expert, arrived at on the basis of a set of attributes (e.g., oscillation amplitude, steady offset, convergence rate, divergency), can be expressed as a set of linguistic rules. These rules can be conveniently translated into fuzzy knowledge. Statements relating attribute characteristics and tuning actions can be expressed as decision tables, with membership functions defined for the control actions. In this manner, tuning actions can be determined by the fuzzy controller using the attribute information on a joint response as provided by a servo expert in level 2.

■ SYSTEM DEVELOPMENT

To explore and illustrate the hierarchical knowledge-based control structure discussed previously a control system has been developed for a two-degree-of-freedom robot. This system consists of a robot simulator, two servo experts for the joints of the robot, and a fuzzy controller to generate the tuning commands for the PID controllers of the joints. The system has been developed on a SUN-3/60 workstation and transferred onto a high-speed SUN SPARCSTATION for subsequent demonstrations. The robot simulator was a C program developed, compiled, and stored as a separate UNIX process. The servo experts have been developed using the commercially available MUSE AI toolkit (MUSE, 1987) separately. Rules and other unstructured programs of the servo experts have been written in PopTalk and the structured code of the knowledge-based controller has been developed using the editor-tool facility of MUSE. The fuzzy controller has been developed independently and has been integrated subsequently with the servo experts in the form of a set of decision tables. Real-time communication among processes has been achieved using UNIX Socket capability. Channel objects and Stream objects of MUSE, and additional interface programs were written in PopTalk.

The specific application discussed in this section does not cover all aspects of the knowledge-based control structure in detail. The purpose here is to illustrate the typical steps that are followed in developing a knowledge-based control system of the specific type by using a relatively simple example, and to show the feasibility of the technique. In a practical application the controller would be developed on a host computer such as the SUN workstation used in the present application, then tested using simulated data, and finally loaded into an applications computer (target machine) for testing the controller with real data. This last step, which involves routine procedures of prototyping a controller, is crucial but not described here.

Servo Experts
Each joint has an associated servo expert. Because these servo experts have identical structures, the approach that has been taken is to first develop one servo expert and, after testing it with data inputs, simply to duplicate it for the second joint of the robot. A servo expert is a knowledge system. It consists of a knowledge source, containing a rule system of the forward production type, and several databases. It can interact with other databases, including notice boards, and user-generated programs written in PopTalk.

The MUSE AI Toolkit
A complete description of the MUSE AI Toolkit may be found in the user manuals (MUSE, 1987). This introduction provides an overview of some of the characteristics of the toolkit, thus giving continuity to this chapter.

MUSE provides a flexible environment for developing knowledge-based applications. Due to this flexibility, a variety of application structures can be developed using several forms of knowledge representations. Either FPS or backward chaining systems (BCS) may be used. What is relevant in the present application is that MUSE

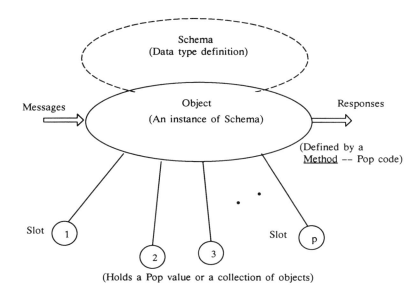

Figure 6 A typical object.

can be used to develop a knowledge system consisting of a knowledge base of forward production rules, several databases, and an inference engine. Conveniently, this is the structure chosen for the proposed servo expert.

MUSE applications are developed by first building structures called objects, and then interconnecting these modules, subject to some structural constraints. An object contains data slots as well as instructions. In a conventional program instructions and data structures are integrated together over the entire program. Hence, even a small change to a data structure could make the program nonfunctional. In object-oriented programs the data structure of any object can be modified without affecting the other objects in the program.

The structure of a MUSE "object" is shown in Figure 6. An object contains a set of slots, and these slots can hold one or more other objects. This results in a tree structure which terminates (at a leaf node) with slots containing unstructured PopTalk program modules or data values. An object can respond to messages. A response will depend on the *method* associated with the message for that particular type of object. Methods are programmed by the user in PopTalk, at the stage of defining the schemes of the objects. An object has a *schema* associated with it. A schema is a data type definition, and it is the structural skeleton (template) from which any number of objects of the type of that particular schema can be produced. In a MUSE application, knowledge is stored in a special type of object called a *knowledge source.* The structure of a knowledge source object

168 INTELLIGENT CONTROL: FUZZY LOGIC APPLICATIONS

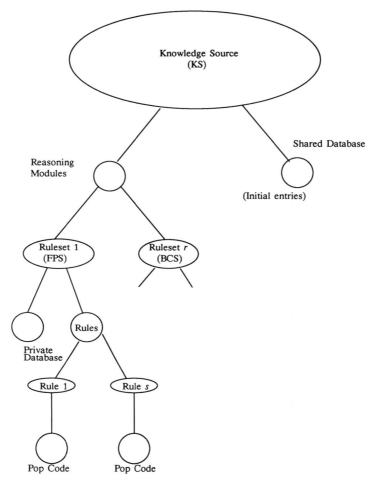

Figure 7 A knowledge resource object.

is shown in Figure 7. It consists of a shared database and a "reasoning modules" slot. The reasoning modules slot can contain one or more rulesets. A ruleset object has a private database that is not visible outside the ruleset, and a *rules slot*. The rules slot has a collection of rule objects. Each rule object has a slot into which an unstructured PopTalk code can be programmed that defines the particular rule in the ruleset of the knowledge base. If the ruleset is of the forward production rules (FPR) type, it corresponds to a forward production system (FPS). Backward chaining system-type rulesets are also available with MUSE, but are not used in the present application. A *notice board* is a database that can store information that will be visible to other objects. The structure of a notice board is

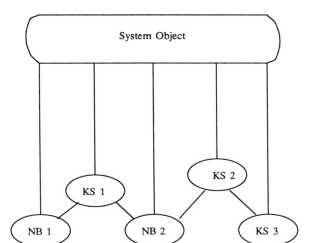

* Structured Editor

* EMACS (Unstructured) Editor

* PopTalk (Compile, Run) Pane

* Browser (Run-Time) Pane

Figure 8 A typical MUSE application.

similar to that of a knowledge source, except that a notice board is a passive object and does not contain rulesets.

A typical MUSE application is built by starting with a core object (handle) known as *system object*, and developing and attaching knowledge sources and notice boards to the system object. This is schematically represented in Figure 8. Object manipulation may be done using the structured editor that is available with MUSE. Rules, methods, library programs, and other unstructured programs are developed in PopTalk using the EMACS editor. An *agenda* provides a scheduling mechanism for the execution of the knowledge sources in a given application.

Three other types of objects important in the present application are demons, data channels, and data streams. A *demon* provides a general mechanism for monitoring objects. It can monitor creation, deletion, and updating of slots in an object, and can perform an action, as programmed in PopTalk, depending on the monitored condition. The action can be made either just prior to the change in the monitored object (pre-update demon) or after the monitored object has changed (post-update demon). A *data channel* is an object that can be interfaced to an external data source (e.g., a simulator) using the UNIX Socket mechanism. A channel object has an associated demon whose action can be programmed in PopTalk. When the channel receives an item of data, its demon is fired. A stream data channel is a data channel that can receive more than one item of data at a time. A *data stream* is an object that can be linked to a file outside the MUSE process. It can either read data from the file or write data into the file. In this manner channel objects and data stream objects can be used for linking MUSE processes with other external processes.

■ SERVO EXPERT DEVELOPMENT

The structure of a servo expert in the present application is shown in Figure 9. Each servo expert consists of a knowledge source, a notice board, and a PopTalk program for interfacing it to an external simulator through a UNIX Socket. There is a servo expert for each degree of freedom (joint) of the robot. The notice board contains the specifications used in evaluating a joint response. The knowledge source contains a ruleset that has the rules carrying the necessary intelligence for evaluating a joint response. The inferences made on the joint response are stored as an object in the knowledge-source database. These inferences are displayed during operation and are used by the fuzzy controller for making servo-tuning decisions at the top level of the control hierarchy. The PopTalk interface program associated with a servo expert creates a channel object for receiving joint response error data from the robot simulator through a UNIX Socket interface. The program generates an error object by taking three successive joint error values received by the channel object. The program is also responsible for the real-time transfer of these error objects to the database of its servo expert.

In the present application three specifications are used for evaluating a joint:

1. An error tolerance (e_s)
2. An oscillation amplitude tolerance (a)
3. An acceptable rate of error convergence (λ)

Six rules are used in the rule base of each servo expert to test five attributes of a joint response. The five attributes considered are

1. Accuracy
2. Oscillations
3. Speed of response (error convergence)
4. Divergence
5. Steady offset

The criteria that are used in the testing are illustrated in Figure 10. The first rule checks whether the joint response is accurate. The

KNOWLEDGE-BASED CONTROL OF ROBOTS   171

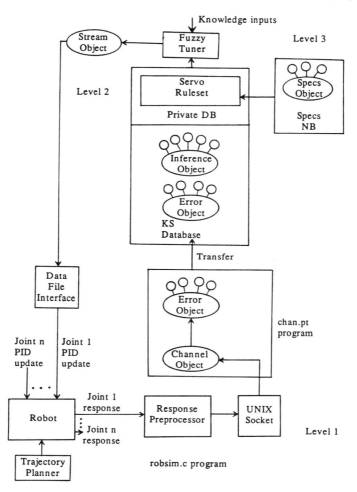

Figure 9   The structure of a servo expert.

second rule checks for unsatisfactory oscillations in the error response. The third rule monitors the speed of convergence of the joint error. The fourth rule checks for steadily diverging unstable behavior in the error response. The fifth rule checks for existence of a steady offset in the joint error. The final rule is needed for continuing the rule search when none of the previous five rules are fired in a given step, and its rule is the link for expanding the rule base. For example, the PopTalk code of the "accuracy" rule is shown in Figure 11.

Because a servo expert should not be expected to make inferences on the basis of a single data sample from a process, a preprocessor (filter) is used within the robot simulator to select the response error values for transmittal to the servo expert. First, an

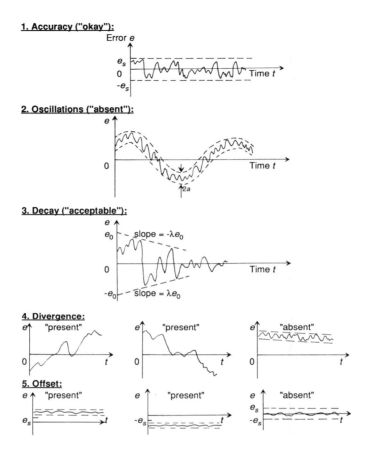

Figure 10 Some attributes of a joint response.

error monitoring period is chosen as an integral multiple of the sampling period of the robot simulator. This monitoring period determines the bandwidth of the knowledge-based controller. The preprocessor observes the joint error response over each monitoring period, and determines the alternate maximum and minimum values of error within successive error monitoring periods. These error values are sent sequentially to the channel object of the servo expert. The fuzzy controller has the ability to modify the monitoring speed (monitoring period) depending on the experience gained during operation.

The PopTalk servo expert program developed for the present application was first tested by injecting error values through an error notice board included during the development stage, and observing the inferences made by the reasoning mechanism. In every test correct inferences have been made by the servo expert. Admittedly,

```
global 4 3 desilva 120388
~ object FPRuleset pid_rules1 {
  desilva 120388 | Joint 1 response testing rules.
~~  Comment
       Rules that make inferences on
       the joint 1 response of robot.
~~  rules
~       collection FPR {
~         object FPR okay_rule1 {
          desilva 120388 | Accurate response.
~~        Comment
             Checks whether the joint 1
             response is accurate.
~~        Source
             if
                there is an inference I
                    -name "infr1" and
                there is a specs
                    -name "spec1",
                -ess E and
                there is an error ER
                    -name "error1",
                -e0 A where (abs(A) =< E),
                -e1 B where (abs(B) =< E),
                -e2 C where (abs(C) =< E)
             then
                do(
                vars strm1 = {stream:};
                open('data1', 1) -> strm1;
                strm1: flushit;
                strm1: putchar(' 0.00   0.00   0.00');
                strm1: flushit;
                strm1: closeit; )      and
                (printf('Joint 1 response is accurate.
                        \n');) and
                assert {inference I: -accuracy "okay"}
                and
                delete ER from KS
```

Figure 11 PopTalk code of the accuracy rule.

the rule base is not complete. The flexibility of a servo expert lies in the capacity to conveniently add new rules and delete unnecessary rules, as further experience is gained and expert advice is available, without having to modify other parts of the overall system. For example, a rule may be added to check for the presence of *reset windup* (integral windup) and to suppress the integral control action on that basis. After testing the first servo expert, the error notice board has been deleted and the servo expert has been duplicated for the second joint of the robot. Subsequently, the new servo expert has been modified to suit the relevant joint.

Robot Simulator

A program has been written in C to simulate a robot. Because nonlinear feedback was used at the servo level to decouple and linearize the robot in the present control structure, it was not necessary to program a nonlinear model of the robot again. The response of the

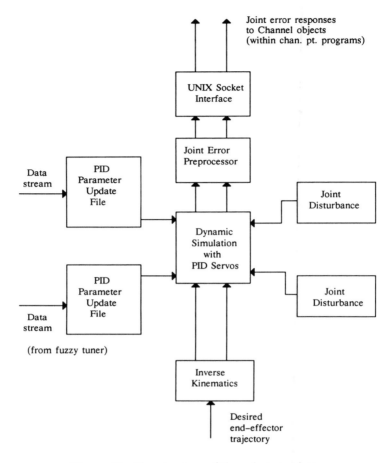

Figure 12 The structure of the robot simulator.

robot with decoupling control and PID servos can be represented by a set of simple oscillators with PID control and unknown disturbances. Hence, the simplifying approach used here has been to program this latter system, with acceleration-type disturbances provided by a set of user-defined functions. Various types of disturbances, including random, pulse, step, periodic, and steady offset, can be easily included in this way. In an actual application the recursive relations (de Silva and MacFarlane, 1989b) could be used to implement a high-bandwidth decoupling and linearizing controller at the lowest level of the control structure. The present simulation takes for granted the effectiveness of such a control algorithm.

The structure of the robot simulator is shown in Figure 12. The *inverse kinematics* program, written in C, computes the desired joint trajectories corresponding to a desired end-effector trajectory programmed into it. The desired joint trajectories are then supplied as

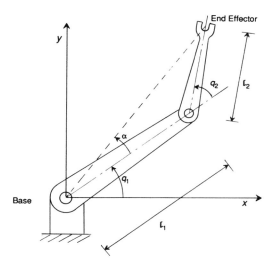

Figure 13 A two-link planar robot.

the reference inputs to the joints of the simulated robot. The equations used in the inverse-kinematics computation are given in the next section. Additional C program units have been provided in the simulator for generating disturbance inputs to the joints. A preprocessor (a C program module) observes the actual joint error responses over each successive error monitoring period, and alternatively picks maximum and minimum values (algebraic) of joint error within each period. These values are then sequentially transmitted to the Channel object of the particular servo expert, through a UNIX Socket interface. Each Channel object is created by a separate PopTalk interface program. The same program constructs error objects from channel data, and transmits the error objects to an appropriate servo expert. The fuzzy controller sends PID parameter updates for each joint into separate data files. These data files are momentarily opened by the robot simulator during each monitoring period to read in the contents. These files are closed immediately after reading the data so that they would be ready to receive further updates from the fuzzy controller.

Inverse-Kinematics Model

The inverse-kinematics problem for a robot is described by de Silva and MacFarlane (1989b). In this section, the equations which are used in the computation of the inverse kinematics for the two-degree-of-freedom robot are given.

Consider the two-link robot shown in Figure 13, which executes planar motions in the Cartesian frame (x, y). The position of the end-effector is given by the vector

$$\mathbf{r} = \begin{bmatrix} r_x \\ r_y \end{bmatrix} \quad (3)$$

It is seen that the direct-kinematics problem is governed by the two equations:

$$r_x = \ell_1 \cos q_1 + \ell_2 \cos(q_1 + q_2) \quad (4)$$

$$r_y = \ell_1 \sin q_1 + \ell_2 \sin(q_1 + q_2) \quad (5)$$

where ℓ_i are the link lengths and q_i are the joint coordinates, as shown in Figure 13.

The equations for inverse kinematics are obtained by simply applying the well-known cosine and sine formulas to the triangle formed by the links of the manipulator. Specifically, we get:

$$q_2 = \cos^{-1}[(r_x^2 + r_y^2 - \ell_1^2 - \ell_2^2)/(2\ell_1 \ell_2)] \quad (6)$$

$$q_1 = \tan^{-1}(r_y / r_x) - \alpha \quad (7)$$

where angle α is as shown in the figure, and is given by:

$$\alpha = \sin^{-1}[\ell_2 \sin q_2 / \sqrt{r_x^2 + r_y^2}] \quad (8)$$

In order to avoid nonunique solutions in the inverse-kinematics problem, the joint angles will be restricted to $0 < q_1 < 90°$ and $0 < q_2 < 180°$.

Fuzzy Controller

The next stage of the present application was the development of the fuzzy controller. First, expert knowledge on tuning a PID servo has been expressed as a set of linguistic statements. Such statements (rules) will necessarily contain fuzzy quantities. Next, membership functions were established for the fuzzy quantities which are present in the fuzzy rules. Each fuzzy rule has been expressed as a fuzzy relation, in tabular form, by incorporating the membership functions into it through fuzzy logic connectives (e.g., IF-THEN implication and OR). The resulting set of tables forms the rule base of the fuzzy tuner. Rule matching can be accomplished by on-line application of the compositional rule of inference to the fuzzy rule base, if so desired, using the inferences made by the servo experts as the context. This on-line application of the compositional rule is generally time consuming, and hence it may considerably reduce the speed of servo tuning. Because the universe of discourse

If the error response is not oscillatory *then* do not change the proportional *and* derivative control parameters;
or if the error response is moderately oscillatory *then* make a small decrease in the proportional gain *and* a small increase in the derivative time constant;
or if the error response is highly oscillatory *then* make a large decrease in the proportional gain *and* a large increase in the derivative time constant

If the error response is fast enough *then* do not change the proportional *and* derivative control parameters;
or *if* the error response is not fast enough *then* make a small increase in the proportional gain *and* a small increase in the derivative time constant.

If the error response does not steadily diverge *then* do not change the proportional *and* integral *and* derivative control parameters; *or if* the error response steadily diverges *then* slightly decrease the proportional gain *and* slightly decrease the integral rate *and* make a large increase in the derivative time constant.

If the error response does not have an offset *then* do not change the proportional *and* integral control parameters;
or if the error response has an offset *then* slightly increase the proportional gain *and* make a large increase in the integral rate.

Figure 14 Linguistic fuzzy rules for tuning a joint servo of a robot.

of the servo-expert inferences (context) is discrete and finite in the present application, and because the universe of the tuning actions is also discrete and finite (both classes of universes having low cardinality), computational efficiency can be significantly improved by applying the compositional rule of inference off line. This preprocessing will generate a decision table for fuzzy tuning of each joint servo. Fuzzy decision tables developed in this manner can then be directly integrated with the servo experts to perform PID tuning on line during process operation. The steps of developing a fuzzy decision table for servo tuning are explained here. For brevity, the procedure is given in detail for one rule in the fuzzy ruleset, and only the final results are given for the remaining rules.

Consider the set of linguistic statements given in Figure 14. These linguistic rules reflect the actions of a human expert in tuning a PID servo by observing the error response of the servo. Note the fuzzy quantities such as "highly oscillatory", "fast enough", and "small increase" that are present in these rules. It is clear that what is given in Figure 14 is a set of linguistic fuzzy rules for servo tuning. Of course, more rules can be added, and the resolution of various fuzzy quantities can be increased to improve tuning accuracy, but for the purpose of the present demonstration, this ruleset is adequate. Now, using the notation given in Figure 15, we can express the fuzzy tuning rules in the condensed form shown in Figure 16.

OSC	=	Oscillations in the error response
RSP	=	Speed of response (decay) of the error
DIV	=	Divergence of the error response
OFF	=	Offset in the error response
OKY	=	Satisfactory
MOD	=	Moderately unsatisfactory
HIG	=	Highly unsatisfactory
NOK	=	Unsatisfactory
DP	=	Change (relative) of the proportional gain
DI	=	Change (relative) of the integral rate
DD	=	Change (relative) of the derivative time constant
NH	=	Negative high (magnitude)
NL	=	Negative low (magnitude)
NC	=	No change
PL	=	Positive low
PH	=	Positive high

Figure 15 The notation used for the fuzzy quantities.

Next, one must develop membership functions for the fuzzy conditions OSC, RSP, DIV, and OFF and for the fuzzy actions DP, DI, and DD. Several methods for estimating such membership functions are known (de Silva and MacFarlane, 1989b) (see Chapter 3). For our purposes, an approximate set of membership functions would suffice, based on intuition. Generally, there is freedom to choose the resolution of each fuzzy quantity. In the present application, however, because inferences made by a servo expert are constrained to a discrete and finite set, the resolution of the condition variables would be fixed. For example, because the variable OSC can take one of three fuzzy values (OKY, MOD, and HIG), the membership function of OKY, for instance, should be assigned a universe having only three elements, each element being representative of one of the three fuzzy values. For action variables, however, this restriction of resolution is not necessary. For example, even though the action variable DP can assume one of four fuzzy values (NH, NL, NC, PL),

		If	OSC	=	OKY	then	DP	=	NC
						and	DD	=	NC
or	If		OSC	=	MOD	then	DP	=	NL
						and	DD	=	PL
or	If		OSC	=	HIG	then	DP	=	NH
						and	DD	=	PH
		If	RSP	=	OKY	then	DP	=	NC
						and	DD	=	PL
		If	DIV	=	OKY	then	DP	=	NC
						and	DI	=	NC
						and	DD	=	NC
or	If		DIV	=	NOK	then	DP	=	NL
						and	DI	=	NL
						and	DD	=	PH
		If	OFF	=	OKY	then	DP	=	NC
						and	DI	=	NC
or	If		OFF	=	NOK	then	DP	=	PL
						and	DI	=	PH

Figure 16 Condensed form of the linguistic fuzzy rules.

the universe of, e.g., NH, may contain more than four elements. However, in the interest of consistency and simplicity we have decided to use the same restriction in resolution for both action variables and condition variables. Specifically, the cardinality of the universe of discourse of a fuzzy quantity has been taken to be equal to the number of fuzzy states which can be assumed by the particular attribute. Accordingly, the universe of NH has been assigned a cardinality of 4, and so on.

In an application such as ours, actual numerical values given to the elements in a universe of a fuzzy quantity are of significance only in a relative sense. An appropriate physical meaning should be attached to each value; however. For instance, in defining DP, the numerical value −2 has been chosen to represent a negative high change in the proportional gain. This choice is compatible with the choice of −1 to represent a negative low change. Also, such numerical values may be scaled and converted as necessary to achieve physical and dimensional compatibility.

The chosen membership functions are shown in Tables 2 and 3. Note that a membership grade of unity has been assigned, as desired, to the representative value of each fuzzy quantity. Fuzziness was introduced by assigning uniformly decreasing membership grades, starting with a low grade (0.2 or 0.1), to the remaining element values.

TABLE 2
MEMBERSHIP FUNCTIONS OF THE CONDITION VARIABLES

OSC

	0	1	2
OKY	1.0	0.2	0.1
MOD	0.2	1.0	0.2
HIG	0.1	0.2	1.0

RSP

	0	1
OKY	1.0	0.2
NOK	0.2	1.0

DIV

	0	1
OKY	1.0	0.1
NOK	0.1	1.0

OFF

	0	1
OKY	1.0	0.2
NOK	0.2	1.0

TABLE 3
MEMBERSHIP FUNCTIONS OF THE ACTION VARIABLES

DP

	−2	−1	0	1
NH	1.0	0.2	0.1	0.0
NL	0.2	1.0	0.2	0.1
NC	0.1	0.2	1.0	0.2
PL	0.0	0.1	0.2	1.0

DI

	−1	0	2
NL	1.0	0.2	0.0
NC	0.2	1.0	0.1
PH	0.0	0.1	0.0

DD

	−1	0	1	2
NL	1.0	0.2	0.1	0.0
NC	0.2	1.0	0.2	0.1
PL	0.1	0.2	1.0	0.2
PH	0.0	0.1	0.2	1.0

Next we shall show the development of a fuzzy relation table for a group of fuzzy rules. Table 3 illustrates the steps associated with this development for the rules relating the condition OSC and the action DP. Note that there are three such rules in the rule base. For example, consider the rule:

$$\text{IF OSC = OKY THEN DP = NC}$$

To construct its fuzzy relation table R_1, we take the membership function of OKY from Table 2 and the membership function of NC from Table 3. Next, in view of the fact that fuzzy implication is a *min* operation on membership grades, we form the Cartesian product space of these two membership function vectors and assign the lower value of each pair of membership grades to the corresponding location in the Cartesian space. The remaining two relation tables R_2 and R_3 were obtained in a similar fashion. The three rules are connected by fuzzy OR connectives. Hence, the composite relation table R has been obtained by combining R_1, R_2, and R_3 through a *max* operation: take the largest value of each triad of membership grades.

The final step in the development of the fuzzy tuning controller is the establishment of the decision table. To explain this development, suppose that a servo expert infers the presence of moderate oscillations in the joint error response. Generally, this is a fuzzy inference, and its membership function is given by:

$$\mu_{\text{OSC}} = [0.2, 1.0, 0.2]$$

Alternatively, the inference may be assumed crisp, with the membership function:

$$\mu_{\text{OSC}} = [0, 1, 0]$$

In either case, this context must be matched with the rule base R obtained in Table 4. This has been accomplished by applying the compositional rule of inference. Recall that this is a *sup* (or, *max* in the discrete case) of *min* operation. Specifically, we compare the vector with each column of R, take the lower value in each pair of compared elements, and then take the largest of the three elements thus obtained. It can be verified easily that by this procedure we obtain:

$$\mu_{\text{DP}} = [0.2, 1.0, 0.2, 0.2]$$

as the membership function of the action on the proportional gain, corresponding to the inferred context. This result now must be

TABLE 4
DEVELOPMENT OF A FUZZY RELATION TABLE (FOR OSC → DP)

R_1: IF OSC = OKY THEN DP = NC

		DP			
		-2	-1	0	1
	0	0.1	0.2	1.0	0.2
OSC	1	0.1	0.2	0.2	0.2
	2	0.1	0.1	0.1	0.1

R_2: IF OSC = MOD THEN DP = NL

		DP			
		-2	-1	0	1
	0	0.2	0.2	0.2	0.1
OSC	1	0.2	1.0	0.2	0.1
	2	0.2	0.2	0.2	0.1

R_3: IF OSC = HIG THEN DP = NH

		DP			
		-2	-1	0	1
	0	0.1	0.1	0.1	0.0
OSC	1	0.2	0.2	0.1	0.0
	2	1.0	0.2	0.1	0.0

Composite $R = R_1 \vee R_2 \vee R_3$

		DP			
		-2	-1	0	1
	0	0.2	0.2	1.0	0.2
OSC	1	0.2	1.0	0.2	0.2
	2	1.0	0.2	0.2	0.1

defuzzified in order to obtain a crisp value for the tuning action. The center of gravity (centroid) method was used in this application. Specifically, the elements in the universe (strictly, the support set) of DP have been weighted using the membership grades of the action, and then the average has been taken; thus:

$$\frac{((-2) \times 0.2 + (-1) \times 1.0 + 0 \times 0.2 + 1 \times 0.2)}{(0.2 + 1.0 + 0.2 + 0.2)} = -0.75$$

This value is the entry for the DP action corresponding to the context OCS = MOD in the fuzzy decision table. The fuzzy decision table obtained by following these steps for every rule in the fuzzy rule base is given in Table 5. It has been decided to take no action

TABLE 5
FUZZY DECISION TABLE FOR A JOINT SERVO

Condition	Action		
	DP	DI	DD
OKY	0.0	0.0	0.0
OSC = MOD	−0.75	0.0	0.75
OSC = HIG	−1.33	0.0	1.33
RSP = NOK	−0.36	0.0	0.36
DIV = NOK	−0.55	−0.55	0.91
OFF = NOK	0.25	1.0	0.0

when the conditions are satisfactory, even though an optimal tuning strategy would suggest some other action under that condition. Also, the same decision table has been used for both joints because no discrimination is needed except for proper scaling. The relation used for updating a PID parameter is

$$p_{new} = p_{old} + \Delta p (p_{max} - p_{min}) p_{sen} \qquad (9)$$

in which p denotes the PID parameter. The subscript "*new*" denotes the updated value and "*old*" denotes the previous value. The incremental action taken by the fuzzy controller is denoted by Δp. Upper and lower bounds for a parameter are denoted by the subscripts *max* and *min*. A sensitivity parameter p_{sen} also was introduced by adjusting the sensitivity of tuning, when needed.

Performance Evaluation
This section presents some representative results obtained from the application described previously. The performance of the knowledge-based controller is discussed on the basis of a set of simulation experiments, and some limitations of the present application are noted.

Several simulation experiments have been carried out, using the application described previously in order to evaluate the performance of the proposed knowledge-based controller. In each experiment a comparison was made with the robot performance under conventional control. The conventional controller was assumed to contain the same nonlinear feedback controller as in the knowledge-based controller, but the values of the PID servo parameters were not updated in the conventional control. In other words, the conventional-control results have been obtained by bypassing the servo experts and the fuzzy tuner of the knowledge-based controller. It follows that the performance of the nonlinear feedback controller cannot be evaluated using the present experiments, but errors of the nonlinear feedback controller can be represented by a set of disturbances injected into the joints of the robot. Specifically, the disturbance vector **d** accounts for possible poor performance of the nonlinear feedback controller. Joint disturbance inputs have been provided by separate program units. Mutually dependent or independent disturbances could be injected into the joints in this manner.

Typical Operation
The robot control application was developed on a SUN Microsystems workstation (SUN-3/60). The simulation experiments could be demonstrated either on the same host machine or on a faster workstation (e.g., SUN SPARCSTATION). All programs, including those for the robot simulator and knowledge-based control, have been stored in the same directory *smart-robot*). A typical simulation experiment is carried out as follows.

At the startup of the system, the suntools command is used to access the capability to interact with multiple and overlapping windows on the screen of the workstation. Then the mouse of the workstation can be used as a convenient user-interface device. The application directory (*smart-robot*) is accessed through the current system window. The MUSE reasoning framework is started by using the *editor-tool* command on this window. Now the MUSE window should appear. Next, a console window is opened, using the mouse. The robot simulator is run in a UNIX shell through this window (the simulator window).

At the start, the simulator code (C programs) is compiled through the simulator window, and the executable code is stored in a file named *robsim*. The data files, which receive the PID updates from the fuzzy tuner, are initialized to zero during this step. Next, the control is transferred to the MUSE window through a mouse interaction. The knowledge-based control code is loaded and compiled through this window, using the PopTalk menu. At this stage, the screen of the workstation will appear, as in Figure 17. The left-hand window, which hides the editor-tool window, is the simulator window. The right-hand bottom window is the PopTalk interaction window, and it is used for running the knowledge-based controller.

A simulation experiment is run by first executing the robot simulator using the *robsim* command on the simulator window, and then executing the knowledge-based controller by typing the command on the PopTalk interaction window. During execution, the updated PID values will be printed on the simulator window, and the inferences from the servo experts will be displayed on the PopTalk interaction window. The appearance of the screen at the end of a typical simulation run is shown in Figure 18.

Single Joint Experiments
Initially, a joint of the robot is evaluated separately. The response of the joint has been simulated for a step input, a ramp input, and a

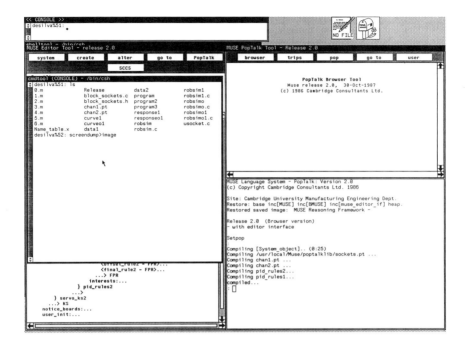

Figure 17 Typical screen view at system startup.

sine input, under both conventional control and knowledge-based control. In these experiments the following parameter values have been used:

Natural frequency of the joint with no control (ω_0) = 100.0 rad/s
Sampling period of the response simulation (Δt) = 2.0 ms
Integral rate of the servo controller (τ_i) = 1.0 s^{-1}
Derivative time constant of the servo controller (τ_d) = 5.0 ms
Bandwidth of the knowledge-based controller = 62.5 Hz
Error convergence rate specification (λ) = 100.0 s^{-1}
Error tolerance specification (e_s) = 0.05 rad
Oscillation amplitude tolerance specification (a) = 0.025 rad

The servo parameter values given above are the values which have been used for the conventional control simulations. They were also the starting values for the knowledge-based control simulations. The step input experiments have been carried out over a duration of 0.1 s, because both conventional and knowledge-based controls gave steady responses within this duration. Double this duration (0.2 s) has been used for the ramp input experiments. A response duration of 1.0 s has been used in the sine input experiments so that at least one full cycle of the input wave could be accommodated.

186 INTELLIGENT CONTROL: FUZZY LOGIC APPLICATIONS

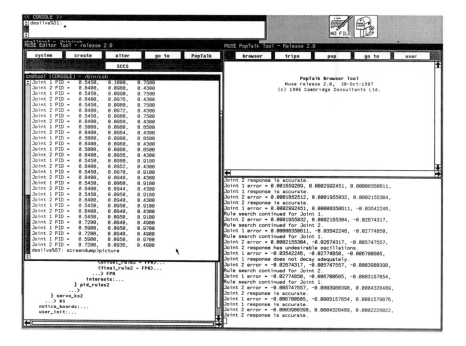

Figure 18 Typical screen view at the end of a simulation.

Results from the step-input tests are shown in Figures 19 and 20. With a knowledge-based controller of bandwidth 62.5 Hz, satisfactory performance has been obtained, but an overshoot is present, as shown in Figure 19(a). By increasing the bandwidth of the knowledge-based controller by a factor of four (to 250 Hz), it has been possible to further improve the performance, completely eliminating the overshoot, as shown in Figure 20 (a).

In the ramp experiments a ramp that reaches a unity magnitude at 0.2 s has been used as the input, to be consistent with the step input tests. The results obtained are shown in Figure 21. The ramp response under knowledge-based control, as shown in Figure 21(a), is quite satisfactory. The knowledge-based controller appears to quickly tune the servo parameters so that accurate tracking of the input ramp is achieved very rapidly. Under conventional control, the ramp response is not as accurate, as shown in Figure 21(b). In particular, the response lags behind the input ramp throughout the duration. Further, a more oscillatory behavior is noticed. On the other hand, the knowledge-based controller has identified the steady offset and has used its integral control action to virtually eliminate this offset.

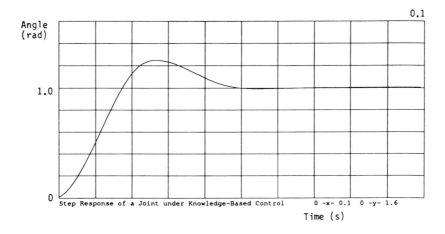

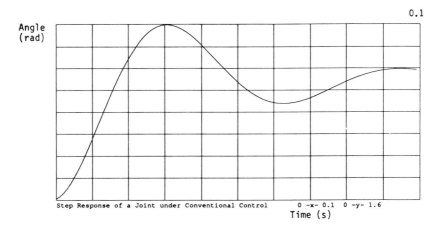

Figure 19 The step response of a joint of the robot (a) under low-bandwidth knowledge-based control and (b) under conventional control.

Results obtained by applying a sine input are shown in Figures 22 and 23. With the knowledge-based controller, the response shown in Figure 22(b) has been obtained for the sinusoidal input of unity amplitude shown in Figures 22(a) and 23(a). When the simulation experiment was repeated under conventional control, the response shown in Figure 23(b) was obtained. It is seen that a significant improvement in the performance has been achieved through the proposed knowledge-based control method.

Seam Tracking Experiments

A robot tracking a right-angular path has been simulated to demonstrate the performance of the proposed controller in seam-tracking

188 INTELLIGENT CONTROL: FUZZY LOGIC APPLICATIONS

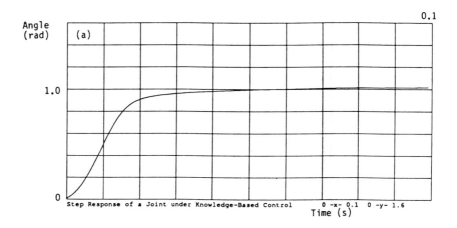

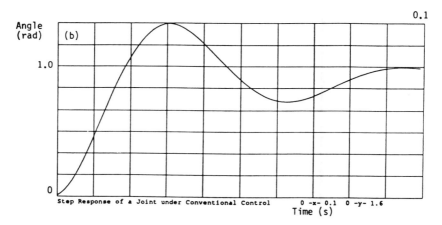

Figure 20 The step response of a joint of the robot (a) under high-bandwidth knowledge-based control (b) under conventional control.

tasks. Several considerations must be taken into account in designing experiments of this type. For example, a suitable time trajectory for tracking the path must be used. Two types of trajectory have been used in these simulations. First, uniformly accelerating and decelerating trajectories of equal duration, without any uniform speed segments, have been used. This would be the case if the capacity of the joint actuators were such that the maximum operating speed is not reached during the task duration, and the task must be completed in minimum time. It has been assumed that the robot comes to rest at the corner of the right angle at a moderate deceleration before changing direction. In the second type of trajectory the end-effector attempts to negotiate the corner much faster. Specifically, in the neighborhood

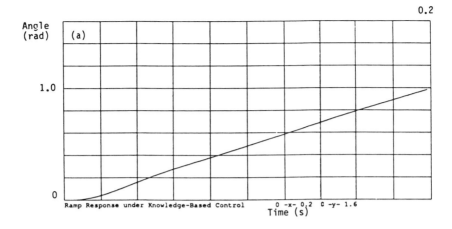

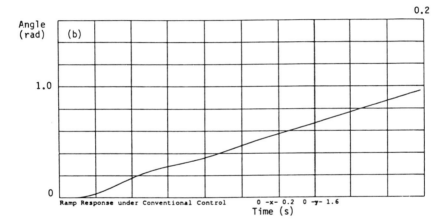

Figure 21 The ramp response of a joint of the robot (a) under knowledge-based control (b) under conventional control.

of the corner, the end-effector is rapidly decelerated so that it comes to rest at the corner. Then the end-effector is rapidly accelerated to its full speed in the orthogonal direction. Uniform speed segments do not present such serious control problems as do accelerating or decelerating segments. For this reason, uniform speed segments have not been included in the seam tracking experiments described here.

Another important consideration lies in the general nature of the robotic task. In some trajectory-following tasks such as arc welding and spray painting it is difficult to start the robot under ideal operating conditions. Hence, an initial position error would be present. Subsequently, uniform operating conditions would be attained, however. Tasks of this category have been simulated by injecting position

190 INTELLIGENT CONTROL: FUZZY LOGIC APPLICATIONS

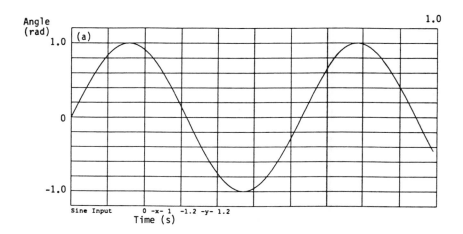

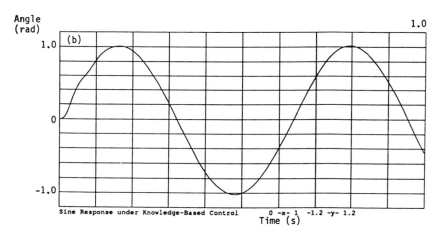

Figure 22 The sine response of a joint of the robot under knowledge-based control (a) The input (b) The response.

errors to both joints of the robot at the beginning of each task. In some other types of task the payload itself might change during operation. Pick-and-place operations, assembly tasks, and interactive tasks for multiple robots, or for a robot and at least one other machine tool (e.g., in a flexible manufacturing cell), would fall into this category. The robot might not stop while the condition change is taking place. Operation under an unsatisfactory nonlinear feedback controller can also lead to a robot behavior similar to what would be present under these conditions. Robot operation under such conditions has been simulated by injecting acceleration pulses to disturb the joints of the robot at different time points during the operation.

KNOWLEDGE-BASED CONTROL OF ROBOTS 191

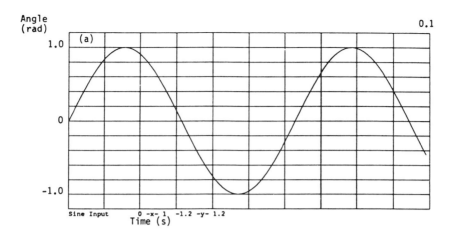

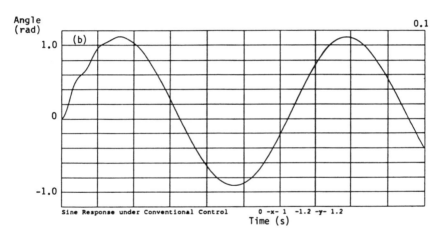

Figure 23 The sine response of a joint of the robot under conventional control. (a) The input. (b) The response.

Natural frequency of joint 1 with no control (ω_1) = 100.0 rad/s
Natural frequency of joint 2 with no control (ω_2) = 150.0 rad/s
Sampling period of the response simulation (Δt) = 8.0 ms
Integral rate of servo controller 1 (r_{i1}) = 0.0 s^{-1}
Integral rate of servo controller 2 (r_{i2}) = 0.0 s^{-1}
Derivative time constant of servo controller 1 (τ_{d1}) = 0.1 ms
Derivative time constant of servo controller 2 (τ_{d2}) = 0.1 ms
Bandwidth of the knowledge-based controller = 62.5 Hz
Error convergence rate specification (λ) = 100.0 s^{-1}
Error tolerance specification for joint 1 (e_{s1}) = 0.01 rad
Error tolerance specification for joint 2 (e_{s2}) = 0.01 rad
Oscillation amplitude tolerance for joint 1 (a_1) = 0.005 rad
Oscillation amplitude tolerance for joint 2 (a_2) = 0.005 rad

In each experiment a right-angular path of dimensions 1.0 × 1.0 m has been tracked in 4.0 s by a robot whose base link is 2.0 m long and whose end-effector link is 1.0 m long.

The robot has been started with errors of magnitude 0.05 rad at the two joints. Note that under the worst circumstances, this can result in an initial position error of 15.0 cm. The end effector has been uniformly accelerated from rest, decelerated to rest at the corner, accelerated again in the orthogonal direction, and decelerated to rest at the end of the path, during four successive periods of 1.0 s along the trajectory. Figure 24(a) shows the response of the robot under knowledge-based control. The controller first corrects the initial error and maintains the accuracy of the response thereafter. The response shown in Figure 24(b) indicates that the conventional controller has been unable to eliminate the position error. It should be mentioned that better performance could have been obtained from the conventional controller by tuning its servo parameters. By the same token, the performance of the knowledge-based controller also could have been improved further by properly selecting the initial parameter values for the joint servos. In these simulations servo parameter values used in the conventional controller are the starting values of the knowledge-based controller. It is not always easy or straightforward to select proper parameters values for a controller; the knowledge-based controller performs well even with an unsatisfactory set of initial parameter values, while the conventional controller does not.

Next, the simulation was repeated, this time with acceleration disturbances of magnitude 0.05 rad/s^2 injected to joint 1 (base joint) of the robot at 0.8, 1.6, 2.4, and 3.2 s, and acceleration disturbances of magnitude 0.1 rad/s^2 injected independently to joint 2 (end-effector joint) at 0.6, 1.4, 2.2, and 3.4 s over the 4.0 s duration of the trajectory. Note that a joint acceleration of magnitude 0.1 rad/s^2 can create a force of 0.3 N on a 1.0 kg mass at the end-effector. The results of this simulation are shown in Figure 25. Again, the knowledge-based controller has been able to compensate for the effects of the disturbances much better than has the conventional controller.

In the next experiment the injected acceleration disturbance sequence was modified in the following way. The two joints have been disturbed simultaneously at the time points 0.8, 1.6, 2.4, and 3.4 s using the pairs (joint 1, joint 2) of acceleration pulses (0.05, 0.1), (−0.05, −0.1), (0.1, 0.05), and (−0.1, −0.05) given in rad/s^2. This simulation may represent, for example, two different pick-and-place operations within the same task. The results of this experiment are given in Figure 26. The response under conventional control appears to deteriorate through each successive disturbance, while the knowledge-based controller makes a good effort to improve the robot response after each disturbance.

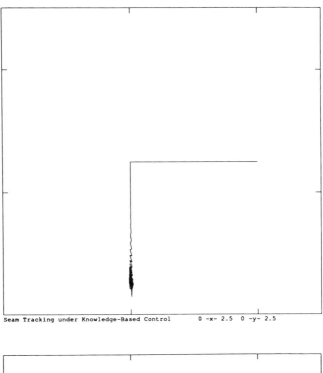
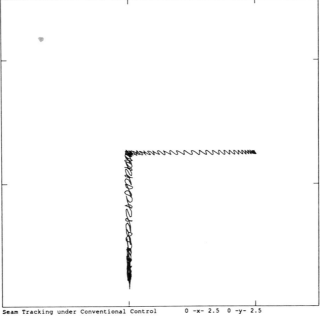

Figure 24 Seam tracking with an initial position error (a) under knowledge-based control and (b) under conventional control.

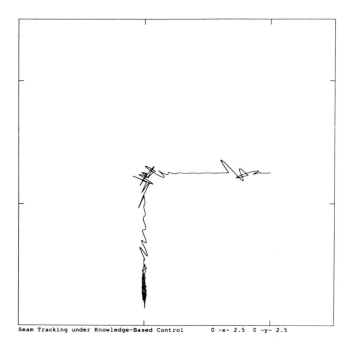

Seam Tracking under Knowledge-Based Control   0 -x- 2.5 0 -y- 2.5

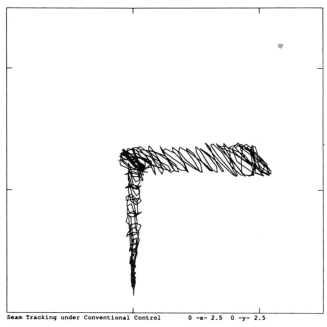

Seam Tracking under Conventional Control   0 -x- 2.5 0 -y- 2.5

Figure 25 Seam tracking subjected to independent joint disturbances (a) under knowledge-based control and (b) under conventional control.

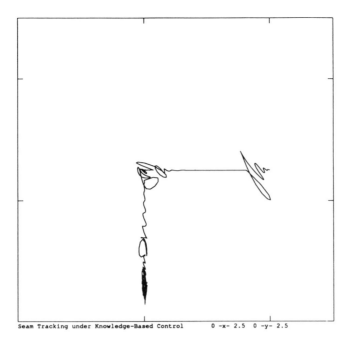

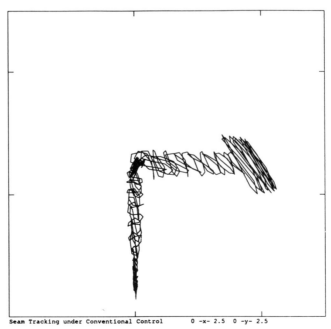

Figure 26 A task with two pick-and-place operations (a) under knowledge-based control (b) under conventional control.

Finally, a faster direction-change operation was simulated under the same sequence of disturbances as in the previous simulation. Specifically, the end-effector has been uniformly accelerated during the first 1.5 s, and rapidly decelerated to rest during the next 0.5 s at the corner. Then, the end-effector has been turned through 90° and rapidly accelerated to its full speed in 0.5 s, and subsequently brought to rest at the end of the path during the final 1.5 s. The results of this experiment are shown in Figure 27. Note the presence of a high-frequency trajectory error in the neighborhood of the corner, due to rapid deceleration and acceleration in that region. The knowledge-based controller seems to cope better with these high-frequency excitations.

Performance and Limitations
The simulation experiments presented here clearly show that the proposed knowledge-based controller is generally superior to a conventional controller which employs a low-level crisp algorithm. However, the question, at what added cost and complexity is this improvement in performance achieved?, remains to be answered. We shall now address some of the relevant issues of the performance and limitations of the present knowledge-based controller, particularly in the context of the application that has been discussed and the results that have been presented here.

The knowledge-based controller quickly identifies important trends or attributes of the robot response. The results show that oscillations, slow error convergence, steadily diverging unstable behavior, and steady offsets in the joint responses have all been identified by the joint servo experts, as expected. Also, the servo tuning decisions made by the fuzzy controller have been found to be satisfactory.

Note that the rule bases used in the servo experts and the linguistic rules in the fuzzy knowledge base are the key to the performance of the overall controller. Once these rules are established, not much effort has been needed to "tune" the knowledge-based controller to achieve a satisfactory performance. In fact, the very first set of trial parameter values has been able to generate good results in the first series of simulations. This shows that not much expertise is required in selecting parameter values in the beginning, and that reasonable guesses seem to work well. For example, the same fuzzy decision table has been used in all the simulations, and it has contained the very first trial values used in its development. Furthermore, the first trial set of servo specifications that were used has resulted in an acceptable performance. Of course, servo specifications must be modified depending on the required stringency and tolerances of control, and on the physical nature of the process. Another encouraging observation has been that the knowledge-based

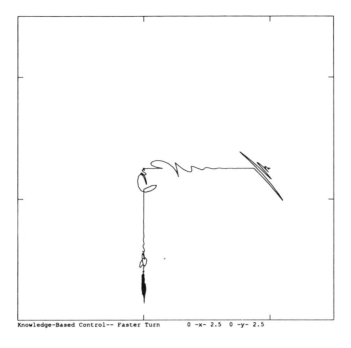

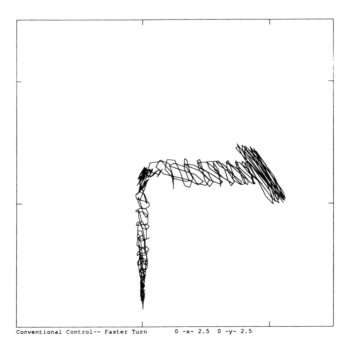

Figure 27 Pick-and-place operations with fast negotiation of a corner (a) under knowledge-based control (b) under conventional control.

controller was quite robust and not very sensitive to the initial values of the servo parameters.

Test inputs of step, ramp, and sine have been used to evaluate the performance of the individual joints. Seam tracking experiments have been used to study the behavior of the overall robot. The trajectory of the end-effector has been specified in seam tracking experiments. The joints of the robot have been expected to perform in a coordinated manner in order to produce the specified end-effector trajectory. In addition to coupling of the servo experts introduced in this manner, another form of dynamic coupling has been introduced through joint disturbances. Coupling errors in a robot, under nonlinear feedback control, can be expressed as a vector of acceleration disturbances at the joints of the robot. Therefore, acceleration disturbances have been injected to the joints to simulate the effects of various tasks such as pick-and-place operations and of coupling errors. Some of these results have shown that the knowledge-based controller somewhat overtunes the servos on the occurrence of the first acceleration disturbance so that subsequent acceleration disturbances would be rapidly compensated. Actually, a better performance could be obtained by modifying the rule base to anticipate such disturbances (and coupling errors) and to take corrective actions accordingly.

Two primary limitations of simulation studies are discussed here. First, the nonlinear feedback control algorithm has not been implemented. Implementation and evaluation of the nonlinear feedback control scheme must be done separately so that any problems due to that control scheme would not overshadow the performance of the knowledge-based controller. Second, an actual physical implementation and testing of the application must be undertaken. This has been accomplished since. Of course, there are several additional issues which must be addressed in a physical implementation. In particular, response sensing, interfacing, signal scaling, and selection of parameter values are all crucial here. Facilities are available from the suppliers of the MUSE AI toolkit for direct loading of the knowledge-based controller (servo experts and the fuzzy tuner) from a SUN workstation into a target machine (typically, a single-board computer) for subsequent interfacing with a process.

Significance

The problem of integrating knowledge-based control into the conventional crisp algorithms of direct digital control has been addressed in this chapter. Although an application of the control approach was demonstrated only for a robotic manipulator, it is known that the general control structure described here is effective in many other types of moderate-to high-bandwidth processes possessing nonlinear and dynamically coupled characteristics. Other physical implementations are described in Chapters 7 and 8. The motivation for the present knowledge-based control approach has come from the observation that knowledge-based control is generally not effective as a direct substitute for conventional methods of standard digital control in moderate- to high-bandwidth processes, and from the fact that linear servo control is normally unsatisfactory in nonlinear and coupled processes of this type. This, combined with the demonstrated fact that human experts can perform controller tuning tasks effectively, has provided the impetus for the approach described in this chapter.

The knowledge-based controller described here has three hierarchical levels. At the lowest level there are conventional servo controllers which are closed around a high-speed nonlinear feedback controller. The second level is the first of two knowledge-based levels in the control structure. This level is somewhat like a knowledge-based preprocessor for measured process information. A servo expert is assigned to each degree of freedom uncoupled by the nonlinear feedback controller. It monitors the response of the degree of freedom and, on the basis of a set of performance specifications, makes inferences as to trends and characteristics of the response in terms of a set of useful attributes. These inferences (attribute values) are passed on to the top level of the hierarchy, the fuzzy control layer. A fuzzy tuner at the top layer uses the inferences made by the servo experts to make tuning decisions. The fuzzy tuner will modify the values of the servo parameters. In a more general context the fuzzy tuner may be assigned many other tasks. It can update the parameter values of the nonlinear feedback controller and even modify the initial performance specifications provided by the user (in level 2). It may perform self-organization and conflict-resolution tasks as well, perhaps employing additional information, including external sensory data and expectational knowledge. For example, the rule bases of the servo experts and the linguistic rules of the fuzzy controller itself could be modified. Furthermore, priorities could be assigned to various degrees of freedom (servos) on the basis of, e.g., measured outputs other than the responses of the degrees of freedom.

It is not feasible to assign a human servo expert to every process that demands a servo expert. In low-bandwidth process control practice it is customary to use general guidelines provided by the supplier of a servo controller for tuning the controller. Typically, tuning is done manually by a trained person (not necessarily an expert). In high-bandwidth processes, particularly those exhibiting nonlinear and coupled characteristics, manual tuning of servos might not provide the required performance accuracy. Conventional crisp algorithms have been used as adaptive controllers to perform tuning at high speed in such situations. However, a major shortcoming of these algorithms is that they are not appropriate for representing the "qualitative", "soft", and "fuzzy" knowledge of a human expert. For example, it would not be possible to directly translate a tuning knowledge base, available as a set of linguistic statements, into a crisp algorithm. Use of a knowledge-based approach (in particular, a fuzzy system) for high-level tuning would be attractive for these reasons.

Other attempts have been made to replace the conventional crisp algorithms of direct digital control by soft knowledge-based methods. Serious drawbacks of such approaches have been recognized in selecting the hierarchical control structure described here. Notably, the control bandwidth can deteriorate significantly by including a soft controller at the servo level, and in real-time DDC it is generally not possible to arbitrarily choose a control bandwidth or to use arbitrarily variable sampling rates. Furthermore, by including a knowledge-based controller into the servo loop, "soft errors" due to noncrisp factors will be directly introduced into drive signals. Also, because any knowledge-based controller requires a finite learning period, and because they are relatively insensitive to initial values, they are more appropriate as tuning controllers. In addition, for a high-bandwidth process, it is not practical to gain experience by manually sequencing the drive signals and manually observing the resulting responses. The knowledge-based control structure described here is attractive in all these respects.

Flexibility of the control structure and the relative ease of further development and modification are further advantages. Typically, one servo expert is developed and tested using suitable inputs, and then it is duplicated for the remaining degrees of freedom of the process. Because all degrees of freedom might not possess similar characteristics, each servo expert has to be examined separately and modified where necessary. Subsequently, parameter values must be assigned for the servo experts by taking into account the characteristics of individual degrees of freedom (e.g., natural frequencies) and the performance requirements (e.g., sensor error tolerances). The development process can be significantly expedited in this manner, particularly for complex and high-order systems. Flexibility of the control structure stems mainly from the ease with which old knowledge can be modified and new knowledge can be added. Because knowledge is present as a set of rules, one is able to simply add or delete appropriate rules. Furthermore, there are built-in mechanisms such as conflict-resolution schemes, to warn about incompatible, redundant, or erroneous rules within the rule base. For example, if it was found that a particular rule was not fired after prolonged operation of the controller, one should re-examine that particular rule for its validity. As another example, if the upper bound of a membership function estimate for a control action variable is not close to unity, then this is an indication that the fuzzy rule base is incomplete. Similarly, multi-modal control inferences may be the result of contradictory rules.

One could argue that once an effective nonlinear feedback controller has been implemented on a process, a satisfactory performance might be achieved by properly choosing the parameter values for the servos and setting them at the beginning of each task, thereby eliminating the need for a knowledge-based tuner. This argument does not hold generally, and even when it holds it is circular in nature. Specifically, some knowledge is needed to choose "proper" values for the servo parameters. The knowledge-based controller can start with a more or less arbitrary set of values and still give good performance. Furthermore, if the servo settings are overspecified, the conventional approach would give good performance, but at the cost of an excessive control effort. The knowledge-based controller will adjust such overspecified conditions as well. Also, it is not possible generally to pick one set of servo parameters which is optimal over a wide range of operating conditions.

Future developments in knowledge-based control should aim at incorporating the well-established merits of crisp algorithmic control in improving the overall effectiveness of a control system. The simple approach of replacing conventional crisp control by direct knowledge-based control in order to make the controller "intelligent" cannot be justified in general. The promise of knowledge-based control lies primarily in its ability to use knowledge available in non-numeric form. Reasoning and associated search procedures in knowledge-based control can be quite slow. Control structures of the form discussed here are important from this point of view because knowledge-based tasks are restricted to slow control zones in these structures. On the other hand, process performance can be improved by increasing the speed of the relevant knowledge-based tasks. Hence, it is important to explore possibilities of reducing the real-time processing overhead of a knowledge-based controller. The decision table approach used in the top level of this control structure, which diverts most of the processing requirements to off line, is one such possibility. These real-time control issues must be carefully studied. When the bandwidths of the knowledge-based structural levels are brought close to that of the crisp-control level, there is the danger of undesirable interactions among control zones in the structure. This is another area that needs careful investigation.

An implementation of self-organizing capabilities, such as on-line modification of the rule bases, could improve the performance of a knowledge-based controller. Enhancement of the knowledge base to detect the level of the coupling errors that are present in a typical case of an imperfect nonlinear feedback controller would be useful. Use of expectational knowledge to improve the performance of the knowledge-based controller is another aspect that deserves exploration. For example, with available knowledge of the nature of a task, it is possible to anticipate some types of external disturbances as well as possible abnormal behavior. This would be the case in a pick-and-place operation of a robot in which impact-type disturbances would arise, or in an assembly operation in which jamming of parts could occur. The rule bases could be appropriately modified to handle such anticipated conditions.

■ SUMMARY

Fuzzy logic has been utilized at several hierarchical levels of a typical robotic control system. Four broad levels of application may be identified: task design, system monitoring (including self-tuning and self-organization), information filtering and preprocessing, and in-loop direct control. Even though the need for fuzzy logic is felt mostly at upper levels of the control system, past applications were mainly concentrated within the lowest level, perhaps driven by convenience rather than necessity. This chapter examined several applications of fuzzy logic in the control of robotic manipulators. Applications were grouped into four hierarchical categories, broadly corresponding to an existing architecture of a robotic control system. Such a classification can be beneficial in ascertaining the appropriateness of fuzzy logic for the specific control task. A high-level implementation using a commercially available AI toolkit was described.

■ PROBLEMS

1. What are advantages of an object-oriented implementation of a robotic workcell? Suppose that each machine tool of the workcell may be considered as a plant with a set of actuators, sensors, and a controller. Sensors are monitored through polling, the information provided by them is interpreted (preprocessed), a control strategy is chosen, and the sensory information is supplied to the controller by the sensor manager. Discuss the use of fuzzy logic in controlling a workcell of this type.

2. Motion control in a constrained direction may result in very large and damaging forces. Similarly, force control in a free (unconstrained) direction can give rise to very large accelerations and decelerations which may lead to unstable behavior. Hybrid force-motion control where force control along one axis and motion control along an orthogonal axis are implemented, may lead to stability, robustness, and task quality problems for these reasons. Could such problems be avoided by using an intelligent control approach? Compare it with impedance control where a relation between motion (velocity) and force is specified, and the controller attempts to reach that relation. A typical application would be in robotics; for example, a robotic hand carrying out a machining operation.

3. Consider vision-based robotics where CCD cameras and associated frame grabber/processor hardware/software, with various levels of image processing. Three situations are considered:
   (a) A high-resolution camera with fast image processing for geometric and motion gauging, which information is then used in low-level control of the robot and other process components.
   (b) A low-resolution camera is used to obtain static images of the work environment, which are then processed off-line for planning the tasks; particularly, trajectory planning of the robot.
   (c) A high-resolution camera is used on-line for determining the "quality" of performance of the robotic system; for example, by analyzing the quality of the processed objects. Here the images are preprocessed to determine the necessary "features" of quality which in turn may be used in a rule-based system to make operational decisions on the robotic system.

   Compare the three cases of robotic vision, identifying the distinguishing characteristics of each case.

4. Consider the rule (knowledge base):

"If an object is within 3.0 ± 0.2 cm of the reference, then snap an image"

Suppose that, at a given instance, the data (context) shows that:

Position of the object = 2.9 cm

The inference of the knowledge-based system would be:

"The object is now ready for imaging"

Compare this with a fuzzy-logic based decision making process where, the rule base

$$R : A \to C$$

and the context

$$\hat{A} = \hat{a} \text{ with membership } \mu \text{ in } A$$

which result in an inference $\hat{C}$ given by

$$\mu_{\hat{C}} = min(\mu, \mu_R) \le \mu$$

5. Consider the production workcell that is schematically shown in Figure 28. There are sensors, actuators, controllers, processors, and the like in the workcell. The hardware configuration of the system, for component interfacing, is shown in Figure 29. The workcell is flexible in the sense that it is programmable and reconfigurable. Also, it is intelligent in the sense that an on-line knowledge-based system is used for monitoring and supervisory control of the workcell. Sketch a suitable hierarchical architecture for the control system of this workcell.

6. A massaging robot has a moderately firm and elastic rotary wheel as the end-effector. The rotating speed ω of the motor that drives the massaging wheel is not constant and changes with the load torque τ. The normal force F, along the axis of rotation of the wheel, is adjustable. The trajectory of massage (both path and associated speed) is planned according to the needs and characteristics of the customer. Note that due to rubbing friction (assume a Coulomb friction model) the load torque τ depends on F. Also, due to viscoelastic type dissipation, τ also depends on the rotary speed ω and trajectory speed v.

One way to adjust F during massaging would be to use an impedance control technique, where a function that relates F to, say, the trajectory speed is specified as the required performance that is sought by the controller. Alternatively, a knowledge-based control approach may be employed. Outline such a technique and indicate the key considerations that are involved.

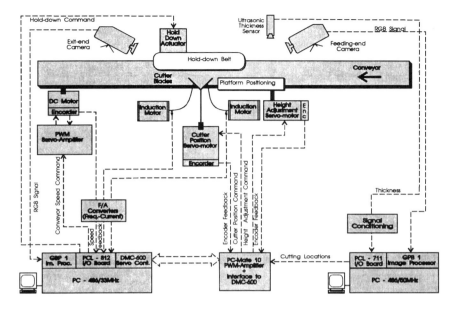

Figure 28 Schematic representation of a production workcell.

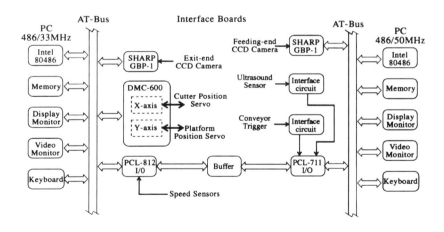

Figure 29 The configuration of component interfacing of the workcell.

7. Robotic tasks may involve combinations of gross motion and fine manipulation. For example, in robotic microsurgery, moving a surgical appliance from point A to point B may involve straightforward gross motions and coarse manipulation. But, actual

performance of a microsurgery will necessitate delicate and dexterous, fine manipulation. A variable-resolution controller might be suitable in such applications. Specifically, coarse-resolution controls will be used for gross manipulation, and fine resolution controls for micro-manipulation. Could fuzzy logic control be exploited in such a robotic application?

8. How is the control of a flexible manufacturing (production) system, (or, facility control), different from that of a workcell or a machine tool?

    Discuss main considerations in the development of an intelligent control system for a workcell.

9. Give reasons for the difficulty in implementing a customized control scheme in an industrial robot or machine tool.

    Why is it difficult to build a workcell using machine tools, robots, etc. from different manufacturers?

■ REFERENCES

Asada, H., Kanade, T., and Takeyama, I. (1983). Control of a Direct-Drive Arm, *ASME J. Dyn. Syst. Meas. Control*, Vol. 105, pp. 136–142.

Bagchi, A. and Hatwal, H. (1992). Fuzzy Logic-Based Techniques for Motion Planning of a Robot Manipulator Amongst Unknown Moving Obstacles, *Robotica*, Vol. 10, pp. 563–573.

Behmenburg, C. (1993). Model Reference Adaptive Systems with Fuzzy Logic Controllers, *Proc. 2nd IEEE Conf. Control Applications*, Vancouver, Canada, Vol. 1, pp. 171–176.

Benerjee, S. and Woo, P.Y. (1993). Fuzzy Logic Control of a Robot Manipulator, *Proc. 2nd IEEE Conf. Control Applications*, Vancouver, Canada, Vol. 1, pp. 87–88.

Cotsaftis, M. (1993). Intelligent Control of Complex Robotic Systems, *Proc. IEEE/RSJ Int. Conf. Intelligent Robots and Systems*, Yokohama, Japan, pp. 708–711.

de Silva, C.W. (1984). A Motion Control Scheme for Robotic Manipulators, *Proc. 1984 Canadian CAD/CAM and Robotics Conf.*, Toronto, Vol. 13, pp. 1–7.

de Silva, C.W. (1985). Motion Sensors for Industrial Robots, *Mech. Eng.*, 107, pp. 40–51.

de Silva, C.W. (1986). Advanced Techniques for Robotic Manipulator Control, *Proc. 1986 Int. Cong. Tech. and Tech. Exchange*, Pittsburgh, pp. 148–153.

de Silva, C.W. (1989). *Control Sensors and Actuators* (Prentice-Hall, Englewood Cliffs, NJ.

de Silva, C.W. (1991a). Fuzzy Information and Degree of Resolution Within the Context of a Control Hierarchy, *Proc. IEEE Int. Conf. Industrial Electronics, Control, and Instrumentation*, Kobe, Japan, pp. 1590–1595.

de Silva, C.W. (1991b). An Analytical Framework for Knowledge-Based Tuning of Servo Controllers, *Eng. Appl. Artif. Intelligence*, Vol. 4, pp. 177–189.

de Silva, C.W. (1992). Research Laboratory for Fish Processing Automation, *Int. J. Robotics Comput.-Integrated Manuf.*, Vol. 9, pp. 49–60.

de Silva, C.W. (1993). Hierarchical Preprocessing of Information in Fuzzy Logic Control Applications, *Proc. 2nd IEEE Conf. Control Applications*, Vancouver, Canada, Vol. 1, pp. 457–461.

de Silva, C.W. (1995). Applications of Fuzzy Logic in the Control of Robotic Manipulators, *Fuzzy Sets and Systems*, Vol. 70, No. 2–3, pp. 223–234.

de Silva, C.W., Ching, C.L., and Lawrence, C. (1988). Base Reaction Optimization of Robotic Manipulators for Space Applications, *Proc. 19th Int. Symp. Industrial Robots*, Sydney, Australia, pp. 829–852.

de Silva, C.W. and MacFarlane, A.G.J. (1988). Knowledge-Based Control Structure for Robotic Manipulators, *Proc. IFAC Workshop on Artificial Intelligence in Real-Time Control*, Swansea, U.K., pp. 143–148.

de Silva, C.W. and MacFarlane, A.G.J. (1989a). Knowledge-Based Control Approach for Robotic Manipulators, *Int. J. Control*, Vol. 50, pp. 249–273.

de Silva, C.W. and MacFarlane, A.G.J. (1989b). *Knowledge-Based Control with Application to Robots*, Springer-Verlag, Berlin.

de Silva, C.W. and Van Winssen, J.C. (1987). Least Squares Adaptive Control for Trajectory Following Robots, *ASME J. Dyn. Syst. Meas. Control*, Vol. 109, pp. 104–110.

de Silva C.W. and Wickramarachchi, N. (1992). Use of Intelligent Tuning in a Hierarchical Control System for Automated Fish Processing, *Proc. IFAC Symp. Adaptive Systems in Control and Signal Processing*, Grenoble, France, pp. 139–144.

De Yong, M., Polson, J., Moore, R., Weng C.C., and Lara, J. (1992). Fuzzy and Adaptive Control Simulations for a Walking Machine, *IEEE Control Syst. Mag.*, Vol. 12, pp. 43–50.

Dubois, D. and Prade, H. (1980). *Fuzzy Sets and Systems*, Academic Press, Orlando, FL.

Dubowski, S. and Des Forges, D.T. (1979). The Application of Model-Referenced Adaptive Control to Robotic Manipulators, *ASME J. Dyn. Syst. Meas. Control*, Vol. 101, pp. 193–200.

Hemami, H. and Camana, P.C. (1976). Nonlinear Feedback in Simple Locomotion Systems, *IEEE Trans. Autom. Control*, Vol. 21, pp. 855–860.

Hirota, K., Arari, Y., and Pedrycs, W. (1985). Robot Control Based on Membership and Vagueness, in: Gupta, M.M. et al. (Eds.), *Approximate Reasoning in Expert Systems*, Elsevier, Amsterdam, pp. 621–635.

Hollerbach, J.M. (1980). A Recursive Formulation of Lagrangian Manipulator Dynamics, *Proc. Jt. Autom. Control Conf.*, TP10-B, San Francisco.

Horn, B.K.P. and Raibert, M.H. (1978). Manipulator Control Using the Configuration Space Method, *Ind. Robot*, Vol. 5(2), pp. 69–73.

Inoue, K., Takano M., and Sasaki, K. (1993). Type Collection of Robot Manipulators using Fuzzy Reasoning in Robot Design System, *Proc. IEEE/RSJ Int. Conf. Intelligent Robots and Systems*, Yokohama, Japan, pp. 926–933.

Ji, S.H., Kwon, S.K., Kim, H.T., and Park, M. (1993). Tracking Navigation using Fuzzy Inference and Sonar-Based Obstacle Avoidance, *Proc. IEEE/RSJ Int. Conf. Intelligent Robots and Systems*, Yokohama, Japan, pp. 898–903.

Kahn, M.E. and Roth, B., (1971). The Near-Minimum-Time Control of Open Loop Articulated Kinematic Chains, *ASME J. Dyn. Syst. Meas. Control*, Vol. 93, pp. 164–172.

Kim, S.W. and Lee, J.J. (1993). Inverse Kinematic Solution Based on Fuzzy Logic for Redundant Manipulators, *Proc. IEEE/RSJ Int. Conf. Intelligent Robotics*, Yokohama, Japan, pp. 904–910.

Kornblugh, R.D. (1984). *An Experimental Evaluation of Robotic Manipulator Dynamic Performance under Model Referenced Adaptive Control*, S.M. thesis, Massachusetts Institute of Technology, Cambridge.

Kumbla, K. and Jamshidi, M. (1993). Fuzzy Control of Three Links of a Robotic Manipulator, *Proc. Int. Fuzzy Systems Assoc. World Congress*, Seoul, Korea, Vol. 3, pp. 1410–1413.

Lim, C.M. and Hiyama, T. (1991). Application of Fuzzy Logic Control to a Manipulator, *IEEE Trans. Robotics Autom.*, Vol. 7, pp. 688–691.

Lin, Y.J. and T.S. Lee, T.S. (1993). An Investigation of Fuzzy Logic Control of Flexible Robots, *Robotica*, Vol. 11, pp. 363–371.

Luh, J.Y.S., Walker, M.W., and Paul, R.P.C. (1980). On-Line Computation Scheme for Mechanical Manipulators, *ASME J. Dyn. Syst. Meas. Control*, Vol. 102, pp. 69–76.

Mamdani, E.H. (1977). Application of Fuzzy Logic to Approximate Reasoning using Linguistic Synthesis, *IEEE Trans. Comput.*, Vol. 26, pp. 1182–1191.

MUSE Support System Manual Set, (1987). Cambridge Consultants Ltd., Cambridge, U.K.

Pham, D.T. and Hafeez, K. (1992). Fuzzy Qualitative Model of a Robot Sensor for Locating Three-Dimensional Objects, *Robotica* Vol. 10, pp. 555–562.

Procyk, T.J. and Mamdani, E.H. (1979). A Linguistic Self-Organizing Controller, *Automatica*, Vol. 15, pp. 15–30.

PROTUNER 1100 Instruction Manual Set, (1984). Techmation Inc., Tempe, AZ.

Scharf, E.M. and Mandic, N.J. (1985). The Application of a Fuzzy Controller to the Control of a Multi-Degree-of-Freedom Robot Arm, in: Sugeno, M. (Ed.), *Industrial Applications of Fuzzy Control*, North-Holland, Amsterdam, pp. 41–61.

Slotine, J.J.E. (1985). The Robust Control of Robot Manipulators, *Int. J. Robotics Res.*, Vol. 4, pp. 49–64.

Staugaard A.C. (1987). *Robotics and AI*, Prentice-Hall, Englewood Cliffs, NJ.

Tokumaru, H. and Iwai, Z. (1972). Non-Interacting Control of Nonlinear Multivariable Systems, *Int. J. Control*, Vol. 16, pp. 945–958.

Tong, R.M. (1977). A Control Engineering Review of Fuzzy Systems, *Automatica*, Vol. 13, pp. 559–569.

Uicker, J.J. (1965). On the Dynamic Analysis of Spatial Linkages using 4×4 Matrices, Ph.D. thesis, Northwestern University, Evanston, IL.

Van Brussels, K. and Vastmans, L. (1984). A Compensation Method for the Dynamic Control of Robots, *Proc. Conf. Robotics Research*, MS 84-487, Bethlehem, PA.

Watanabe, K., Tang, J., Nakamura, M., Koga, S., and Fukuda, T. (1993). Mobile Robot Control using Fuzzy-Gaussian Neural Networks, *Proc. IEEE/RSJ Int. Conf. on Intelligent Robots and Systems*, Yokohama, Japan, pp. 919–925.

Whitney, D.E. (1969). Resolved Motion Rate Control of Manipulators and Human Prosthesis, *IEEE Trans. Man-Machine Syst.*, Vol. 10, pp. 47–53.

Wu, C.H. and Paul, R.P. (1982). Resolved Motion Force Control of Robot Manipulators, *IEEE Trans. Syst. Man Cybernetics*, Vol. 12, pp. 266–275.

Xu, Y. and Hechyba, M.C. (1993). Fuzzy Inverse Kinematic Mapping: Rule Generation, Efficiency and Implementation, *Proc. IEEE/RSJ Int. Conf. Intelligent Robotics*, Yokohama, Japan, pp. 911–918.

Young K.K.D. (1978). Controller Design for a Manipulator using Theory of Variable Structure Systems, *IEEE Trans. Syst. Man Cybernetics*, Vol. 8, pp. 101–109.

Zadeh L.A. (1975). The Concept of a Linguistic Variable and Its Application to Approximate *Reasoning, Inf. Sci.*, Vol. 8, pp. 199–357.

Zhou J. and Raju G.V.S. (1993). Fuzzy Rule-Based Approach for Robot Motion Control in the Presence of Obstacles, *Proc. IEEE Int. Conf. Systems, Man and Cybernetics*, Le Touquet, France, Vol. 4, pp. 662–667.

7 SERVO MOTOR TUNING

■ INTRODUCTION

Even though crisp tuning schemes based on test responses, such as the Ziegler-Nichols method (1942), have been available for over half a century, tuning of servo-control systems is still largely an art. However, there is valuable knowledge that is available to us in the form of operator experience in tuning servo systems. Particularly for servo loops with proportional-integral-derivative (PID) control, such knowledge may be presented in the form of linguistic rules containing fuzzy terms such as "fast", "slight", and "poor". This was central to the knowledge-based tuning scheme developed by de Silva and MacFarlane (1989), as described in Chapters 5 and 6. Subsequently, an analytical framework was developed for that class of problems (de Silva, 1991). Details are presented in Chapter 5.

In the application to robots as developed by de Silva and MacFarlane (1989) and described in Chapter 6, the knowledge base consisted of rules that directly related the nature of the process response, when excited by a test input, to the parameters of the PID controller, the tuning parameters. Even though in the context of operator experience it is realistic to gain tuning knowledge by adjusting the tuning parameters and observing the process response, in a complex control system it may not be easy to "learn" the underlying relationships in this manner, particularly because the responses are studied mostly to detect "qualitative" trends. Suppose that a *mapping* exists between the set of direct tuning parameters and a more convenient set of *controller attributes*. These attributes should generally possess a physical interpretation and also a better intuitive appeal for the designer of a control system. The assumption made here is that the learning process becomes less difficult when these controller attributes rather than the original tuning parameters are matched with the process response. The mapping between the controller attributes and the tuning parameters may be fuzzy in general, but is crisp in conventional control practice. Furthermore, response evaluation should be simplified if it is carried out in a "relative sense" with respect to the response of a *reference model*, and not in an "absolute sense" as done by de Silva and MacFarlane (1989) and described in Chapters 5 and 6. In this chapter a knowledge-based tuning procedure for servomotors is presented that possesses these two innovations (de Silva and Barlev, 1992). Specifically, a suitable set of controller attributes are used as the *action variables* in the tuning knowledge base, and the response of a reference model is used to evaluate the performance of the actual process, when subjected to a test input. This tuning scheme is implemented on a commercially available servo system having a lead compensator and an integral controller.

In manual tuning of a controller it is convenient to perform the tuning actions in a sequential manner according to some form of priority. The person who tunes the controller may observe some set of attributes, determine the conditions that occur, prioritize them, and perform the tuning actions sequentially. In many problems the governing relations (rules) of conditions and actions may be interpreted as being uncoupled in the sense that only one condition variable is related to only one action variable in a given rule. In a general and more complex rule base each rule may relate more than one condition variable and more than one action variable. This latter situation gives a coupled rule base. It should be clear that the action variables in a rule can be uncoupled simply by repeating the rule with the same condition part, but providing only one action at a time without sacrificing any accuracy in the knowledge base, assuming that the rules are processed simultaneously during the inference procedure. Uncoupling the condition variables is not as trivial, however.

The computational and developmental advantages of using an uncoupled rule base in fuzzy-logic control (FLC) are tremendous. Generally, however, some accuracy is lost by assuming that the rules in a knowledge base are uncoupled. In view of this it is important to examine the conditions under which this assumption can be made without sacrificing the control accuracy. As an example, consider a fuzzy-logic controller with the coupled rule base:

IF A_1 AND B_1 THEN C_1 AND D_1

ELSE IF A_2 AND B_2 THEN C_2 AND D_2

END IF

The latter part of the chapter investigates the conditions under which this coupled rule base may be expressed as

$$\text{IF } A_1 \text{ THEN } C_1^* \text{ AND } D_1^*$$
$$\text{IF } B_1 \text{ THEN } \overline{C}_1 \text{ AND } \overline{D}_1$$
$$\text{IF } A_2 \text{ THEN } C_2^* \text{ AND } D_2^*$$
$$\text{IF } B_2 \text{ THEN } \overline{C}_2 \text{ AND } \overline{D}_2$$

First, this problem is analytically investigated. Next, experiments are carried out employing the same prototype servo-motor system that was developed before, which incorporates a fuzzy-logic based tuning to illustrate the effective use of an equivalent uncoupled rule base (de Silva, 1994).

■ SYSTEM DEVELOPMENT

The general structure of the servo tuner considered here is shown in Figure 1, and consists of a three-level hierarchy (de Silva and MacFarlane, 1989). When the process is in a quiescent state a test signal is applied to the control system and an identical signal is applied to the reference model. The corresponding responses are transmitted to the *performance evaluator* at the upper level, which corresponds to the *servo expert*. At this level the process response and the desired response, as generated by the reference model, are compared with respect to a set of *performance specifications*. On this basis, a corresponding set of *performance indicators* is determined, which gives the degree of agreement of the two responses. This will form the *context* for the *knowledge-based tuner* at the top level. The knowledge base consists of a set of linguistic fuzzy rules relating the performance indicators to the controller attributes. The inference mechanism used here is the familiar *compositional rule of inference* (Zadeh, 1979; de Silva and MacFarlane, 1989), described in Chapters 3 and 4. The inferences made by the knowledge-based tuner at this level are the tuning decisions, but these are expressed usually as defuzzified updates of the controller attributes. These values are then transformed by means of the *tuner mapping module* into the corresponding updates of the controller parameters. The resulting numerical values are directly used in tuning the servo controller. The present hierarchical system is different from what is described in Chapters 5 and 6 due to the presence of the *reference model* as the "performance specifier" and the *tuner mapping module*,

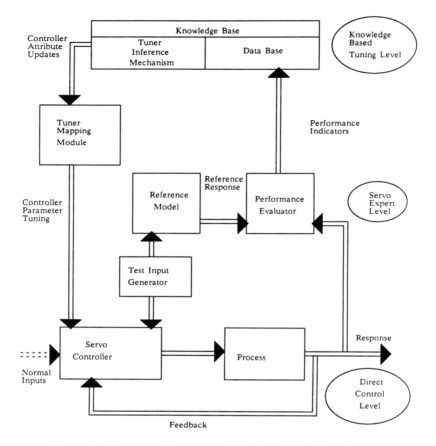

Figure 1 The general structure of the knowledge-based tuner.

which facilitates the use of realistic controller attributes in the knowledge base.

The tuner mapping module can significantly facilitate the development of the knowledge base because physically more meaningful, familiar, and intuitively appealing attributes of the controller may be used as the action variables in the rule base, which are subsequently transformed into the tuning parameters through this module. The knowledge base itself is developed by capturing the past experience, knowledge, and expertise of engineers experienced in tuning the particular class of servo controllers and evaluating this knowledge in conjunction with the principles of control system design. The necessary information may be gained in several ways. For example, the response of the control system to a test input may be determined for a series of values of the controller attributes. Alternatively and more conveniently, a computer model of the control

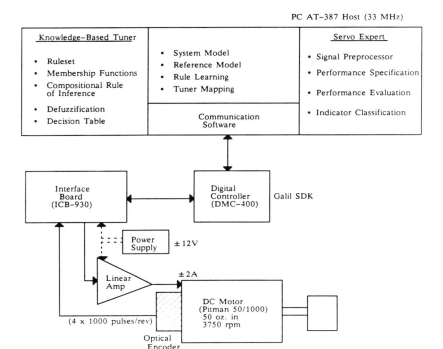

Figure 2 Laboratory implementation of the intelligent tuner.

system may be developed and used to carry out a series of simulation experiments for various values of controller attributes. It should be emphasized that what is used for this purpose is a reasonably accurate model of the actual control system, and is not the "ideal" reference model. The information obtained in this manner may be systematically examined to establish the tuning knowledge. Note that this process of knowledge capturing can be quite fuzzy.

The Experimental System

A laboratory implementation of the intelligent tuner is schematically shown in Figure 2. The system was integrated using a commercially available servo design kit consisting of a direct current (DC) motor with a feedback optical encoder and a linear amplifier (with current feedback) which are interfaced with a programmable digital motion controller that is located in a dedicated host computer (de Silva, 1989. The software modules corresponding to both the servo expert level and the knowledge-based tuning level are also indicated in Figure 2. These are written in FORTRAN, but due to system constraints, some of the communication routines are written in C. A view of the laboratory system is shown in Figure 3.

Figure 3 A view of the laboratory prototype.

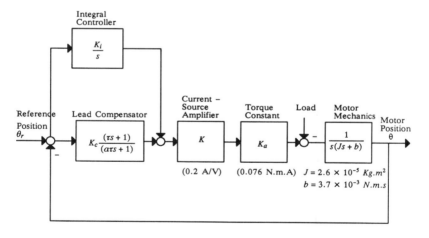

Figure 4 Analytical model of the servo control system.

Model of the Physical System

A model of the servo control system may be used to acquire information about the system behavior and is used, in part, to generate the knowledge base for servo tuning. The model, expressed in the Laplace domain, is given in Figure 4. The system has a current source amplifier which generates a current in proportion to the input voltage from the digital-to-analog converter (DAC) of the digital controller. The permanent-magnet, brushless, DC motor generates a magnetic torque that is proportional to the amplifier current (de Silva, 1989). The digital controller consists of a lead compensator ($\alpha < 1.0$) and an integrator in parallel. In the actual system an 8-bit DAC providing a maximum voltage of 20 V (i.e., 20 V/256 bits) is used to convert the digital command for the motor driver into an appropriate voltage signal. Also, the actual system has an incremental optical encoder for position feedback, and accordingly the motor position would be given in encoder pulses (counts) rather than in radians. (*Note:* 4000 pulses/2π rad.)

Reference Model

A reference model is used for performance specification, and is not the same as the analytical model given above for the physical system. By specifying the desired performance using a dynamic reference model, we are able to guarantee the feasibility of the specifications, at least with respect to the particular reference model. This will provide some buffer against unrealistic specifications. Of course, the reference model may be as complex as necessary, but here, a simple oscillator model is used, having the transfer function

$$G_r(s) = \frac{\theta}{\theta_r} = \frac{\omega_n^2}{(s^2 + 2\zeta\omega_n s + \omega_n^2)} + G_d \qquad (1)$$

in which ω_n and ζ are the *undamped natural frequency* and the *damping ratio* of the reference model. G_d represents both external disturbances (including load) and noise, and may be separated into a mean value, representing the response *offset* and a zero-mean random term representing the *precision* of the response. Using this model, the conventional performance specifications such as *rise time, damping,* and *overshoot* may be conveniently specified (de Silva, 1989).

Knowledge Base

The knowledge base contains a set of rules prescribing what tuning actions should be taken when the actual response of the system deviates from the specifications. The performance deviations are the *condition variables* of the ruleset. Because, generally a "qualitative" comparison between the responses of the actual system and the reference model is made using experience, through a reasoning mechanism of an expert, these deviations are fuzzy quantities. The *action variables* of the rule base are not necessarily the tuning parameters themselves, and are preferably a set of *controller attributes* that are used by control engineers in the conventional design practice. Crisp and unique transformations may exist between these attributes and the tuning parameters. During operation, the context data for the knowledge base are generated as a set of *performance indicators* by

TABLE 1
PARAMETERS DETERMINED BY THE PERFORMANCE EVALUATOR

Performance specification	Rule base notation	Deviation ratio
Rise time (95%)	RT	$1 - RT_m/RT_s$
Damped natural frequency	FD	$1 - FD_s/FD_m$
Damping ratio (average decrement)	DR	$1 - DR_m/DR_s$
Overshoot	OS	$1 - OS_m/OS_s$
Offset	OF	$1 - OF_m/OF_s$
Precision	PR	$1 - PR_m/PR_s$

TABLE 2
CLASSIFICATION OF THE PERFORMANCE DEVIATIONS

Condition	Rule base notation	Category
Overspecified	OV	1
In-specification	IN	2
Marginal	MG	3
Poor	PR	4
Very poor	VP	5

a preprocessor known as the *performance evaluator*, as shown in Figure 1. Once the action variables are determined by the decision-making module to satisfy this context, the actual tuning actions themselves are computed by the *tuner mapping module* shown in Figure 1.

Performance Evaluator

The performance specification is done in terms of the six parameters given in the first column of Table 1. These are computed using the actual responses of the system (s) and the reference model (m) using standard numerical methods, and then the deviation ratios are computed as in the third column of the table. Note that the deviation ratios correspond to the condition variables of the rule base and are defined such that when the value is <0 an overspecification exists, and the worst-case performance is given by 1.0. The numerical values of the deviation ratios are classified into the five categories given in Table 2. These categories are a result of a very nonlinear mapping from the domain of the deviation ratio (Table 1). The category numbers for the six performance parameters, as determined by the performance evaluator, form the context data for the knowledge base.

TABLE 3
DEFINITION OF THE CONTROLLER ATTRIBUTES

Controller attribute	Analytical notation	Rule base notation
Compensator phase lead at regular crossover frequency	ϕ_m	PX
Regular crossover frequency	ω_c	XF
Compensator magnitude (gain) at crossover frequency	g_c	GX
Low frequency of integrator for steady-state accuracy	ω_ℓ	LF

Rule Base of Tuning

The condition variables of the rules are the performance deviations as defined in Table 1. Phase lead (maximum) of the compensator at the crossover frequencies is a measure of stability; specifically, the *phase margin* of the control system. The crossover frequency itself is a measure of the system *bandwidth*, or speed of response. The magnitude (gain) of the compensator at the crossover frequency has an inverse relationship with the process gain and is a measure of the *steady-state accuracy* which also depends on the *external load* on the system. Finally, the frequency ω_ℓ is the point at which the magnitude of the integral controller equals the crossover gain of the compensator. This frequency is a measure of the gain of the integral controller, which in turn determines how fast the steady-state accuracy level is reached in the control system. Because of the qualitative nature of the evaluation process, these four attributes are fuzzy variables in general. Triangular membership functions with discrete support sets (de Silva and MacFarlane, 1989) are assigned to the category marks of these fuzzy variables, as described in previous chapters. The changes in a set of controller attributes form the action variables of the rule base. The four attributes listed in Table 3 are used in the present application. The prescriptions of the necessary adjustments (changes) for them are made primarily through human experience, and hence these changes are fuzzy variables in general.

For each controller attribute five possible fuzzy states are defined, and correspond to fuzzy actions, as given in Table 4. Again, triangular membership functions with discrete support sets are assigned to these fuzzy action states. The tuning rules are established through manual examination and "learning" of the trends of the system response when the controller attributes are changed one at a time. It is thus assumed that each rule has just one condition variable (a performance error) and just one action variable (a change in controller attribute). This significantly simplifies the knowledge base

TABLE 4
FUZZY STATES OF ACTIONS

Action controller attribute	Notation
Negative high	NH
Negative low	NL
No change	NC
Positive low	PL
Positive high	PH

computations. The use of a coupled rule base is studied in the latter part of this chapter.

Tuner Mapping Relations

$$G_c(s) = K_c \frac{(\tau s + 1)}{(\alpha \tau s + 1)} \quad (2)$$

and the integral controller is

$$G_i(s) = \frac{K_i}{s} \quad (3)$$

Mapping of the changes in the four controller attributes (ϕ_m, ω_c, g_c, and ω_ℓ in Table 3) into the corresponding changes in controller parameters (K_c, K_i, τ, and α in Figure 4) is governed by a crisp and unique set of relations in the present application. These equations follow from the conventional procedure of designing a lead compensator in the frequency domain for a servo system (Franklin et al., 1986) and are given by

$$\alpha = \frac{(1 - \sin \phi_m)}{(1 + \sin \phi_m)} \quad (4)$$

$$\tau = \frac{1}{\sqrt{\alpha \omega_c}} \quad (5)$$

$$K_c = \sqrt{\alpha g_c} \quad (6)$$

$$K_i = g_c \omega_\ell \quad (7)$$

Incremental forms of these relations are used in the actual tuning.

RESULTS

First, a rule base is developed using systematic observations of the system response for various levels of controller attributes. Even though computer simulations are used here for convenience, the physical system itself could have been used for that purpose. The four attributes are varied one at a time and the responses are observed. Figure 5 shows some typical data. Step responses with a desired position of 200 encoder counts are shown. Observations that are made using such data include the following:

1. The stability increases and the speed of responses decreases with ϕ_m.
2. The speed of response increases, but the system becomes increasingly oscillatory with ω_c.
3. For a given load the steady-state accuracy improves and the speed of response increases with g_c.
4. The speed at which the steady-state accuracy level is reached increases with ω_ℓ.

One rule is developed for each pair of condition and action variables. For example, the rule relating the *overshoot* (*OS*) and the *compensator phase lead at the crossover frequency* (*PX*) is given below:

IF *OS* IS *VP* THEN *PX* IS *PH*

ELSE IF *OS* IS *PR* THEN *PX* IS *PL*

ELSE IF *OS* IS *MG* THEN *PX* IS *PL*

ELSE IF *OS* IS *IN* THEN *PX* IS *NC*

ELSE IF *OS* IS *OV* THEN *PX* IS *NL*

END IF

Experimental Results

A decision table was developed in the conventional manner (de Silva and MacFarlane, 1989), using the rule base along with discrete membership functions, by applying the *sup-min composition* and defuzzifying through the center of gravity (centroid) method (see Chapter 6). This decision table was utilized in tuning the experimental control system consisting of a commercially available servo motor with a digital controller that contains a *lead compensator* and an *integral controller*. Typical performance of the intelligent tuner is given in Figure 6. The first three results (a) to (c) are for the actual servo system and the fourth result (d) is for the model of the system.

222 INTELLIGENT CONTROL: FUZZY LOGIC APPLICATIONS

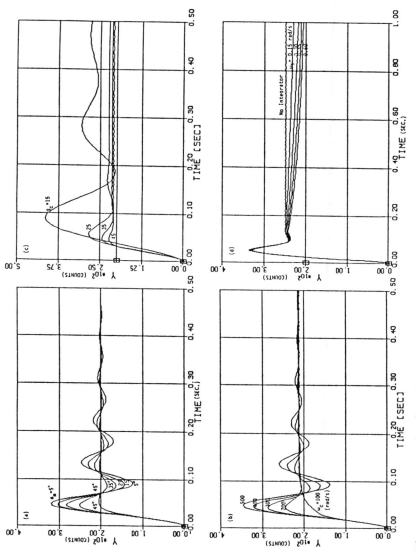

Figure 5 Response data for rule base generation. (a) Effect of maximum phase lead; (b) effect of crossover frequency; (c) effect of compensator gain at crossover; (d) effect of low frequency of integral control.

SERVO MOTOR TUNING 223

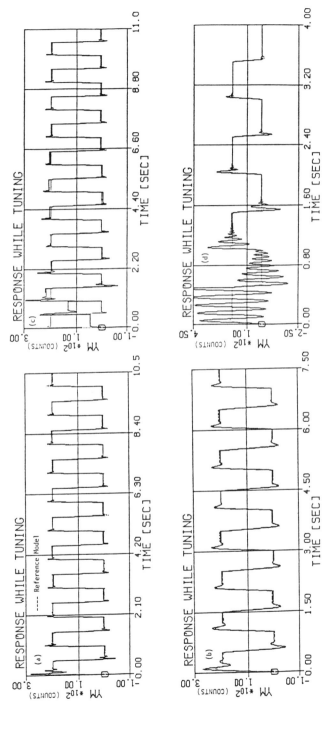

Figure 6 Typical performance of the tuner. (a) Commercial servomotor with oscillations; (b) commercial motor with added inertia; (c) commercial motor with low gain; (d) model with an unstable start.

In each case tuning is progressed until a satisfactory response is achieved. The time required for tuning is an indication of the difficulty of the tuning task. In each of the four experiments the baseline motor was initially disturbed in some manner. Specifically, the physical motor system started with an oscillatory behavior in (a), with an extra inertia coupled to the motor shaft in (b), and with a low control gain in (c). The model of the control system was used in (d), in which the controller values were initially set so that the motor response was unstable. The tuner was able to improve the system performance quite rapidly in all four cases.

■ THEORY OF RULE BASE DECOUPLING

In this section the theoretical basis of rule base decoupling is presented (de Silva, 1994). First, the general fuzzy logic control problem with a coupled rule base is formulated. This represents a multi-degree of freedom decision-making problem. Next, the method of single degree of freedom decision making using a coupled rule base is given. Finally, the assumption of uncoupled rule base is incorporated into the single degree of freedom decision-making problem. On that basis, conditions are established for rule-base decoupling in the problem of single degree of freedom decision making.

Inferencing through a Coupled Rule Base

Consider a knowledge base of fuzzy-logic control, given as a set of linguistics rules in the general form

$$\underset{i,\cdots,a}{\text{ELSE}}[\text{IF } e_1 \text{ is } E_1^i \text{ and } \cdots \text{ and } e_n \text{ is } E_n^a \text{ THEN } c_1 \text{ is } C_1^j \text{ and } \cdots \text{and } c_p \text{ is } C_p^b] \quad (8)$$

with the condition vector $\mathbf{e} \triangleq [e_1, ..., e_n]^T \varepsilon \Re^n$ and the action vector $\mathbf{c} \triangleq [c_1, ..., c_p]^T \varepsilon \Re^p$. Suppose that by monitoring the condition vector $\mathbf{e}$ of the system, a fuzzy context $\hat{\mathbf{E}} \triangleq [\hat{E}_1, \hat{E}_2, ..., \hat{E}_n]^T$ has been established at a given instant. This will provide the membership function $\mu_{\hat{E}}(\mathbf{e})$ of the context. The control decision $\hat{\mathbf{C}} \triangleq [\hat{C}_1, \hat{C}_2, ..., \hat{C}_p]^T$ corresponding to the context $\hat{\mathbf{E}}$ is computed through the use of the *compositional rule of inference* (Zadeh, 1973), as described previously in this book. Specifically, we get

$$\mu_{\hat{\mathbf{C}}}(\mathbf{c}) = \sup_{\mathbf{e}} min[\mu_{\hat{\mathbf{E}}}(\mathbf{e}), \mu_R(\mathbf{e}, \mathbf{c})] \quad (9)$$

Then, crisp control actions $\hat{\mathbf{c}}$ must be computed from fuzzy $\hat{\mathbf{C}}$. In this process first the action membership function, given by Equation (9) is "projected" along each axis c_q and then the *centroid method* is applied to compute the corresponding control action.

Definition
Consider a membership function $\mu(\mathbf{x}, \mathbf{y}): \Re^n \times \Re^p \to [0,1]$. Its *projection* in the $X_i \times Y_j$ subspace is denoted by $Proj[\mu(\mathbf{x}, \mathbf{y})](x_i, y_j): \Re \times \Re \to [0,1]$ and is given by

$$Proj[\mu(\mathbf{x},\mathbf{y})](x_i, y_j) \triangleq \sup_{\substack{\forall x_k \neq x_i \\ \forall y_\ell \neq y_j}} \mu(\mathbf{x},\mathbf{y}) \tag{10}$$

∎

Using this definition of projection, the crisp control actions corresponding to Equation (9) are computed by

$$\hat{c}_q = \frac{\int c_q Proj[\mu_{\tilde{c}}(\mathbf{c})](c_q) dc_q}{\int Proj[\mu_{\tilde{c}}(\mathbf{c})](c_q) dc_q} \qquad q = 1, 2, \cdots, p \tag{11}$$

in which each integration is performed over the support set of the particular membership function.

Arguably, the greatest potential of fuzzy logic control is in applications in which there is a useful knowledge base of manual control, which can be expressed as a set of linguistic rules. Then, tuning and self-organization of a control system, rather than direct low-level control, are better suited for the particular approach (de Silva and MacFarlane, 1989), as discussed in Chapters 5, 6, and 8. Knowledge base acquisition by human operators can be quite effective if only one condition variable and one action variable are considered at a time. Also, control actions determined by the human thought process are often provided one at a time. In complex situations the control actions will be carried out in quick succession and some prioritization of the actions will occur naturally, again depending on factors such as the complexity or urgency of the situation, operator experience, and available expertise. It follows that use of pairs of a single condition variable and a single action variable is quite realistic in fuzzy logic control. For this purpose, however, the specific variables of the knowledge base must be carefully chosen, making particular use of expert experience (de Silva and Barlev, 1992). Now let us address the inference making problem from this particular point of view.

Consider a coupled rule base R, as given by Equation (8), with membership function $\mu_R(\mathbf{e}, \mathbf{c})$. Suppose that a crisp observation $\hat{e}_s$ of the condition variable e_s is available, and it is required to determine a crisp control value $\hat{c}_q$ for the corresponding action variable c_q. First, the projection of μ_R in the $E_s \times C_q$ subspace is determined as

$Proj[\mu_R(\mathbf{e},\mathbf{c})](e_s,c_q)$ using Equation (10)

The crisp condition $\hat{e}_s$ may be expressed as a fuzzy singleton whose membership function is

$$\mu_{\hat{E}_s}(e) \triangleq s(e - \hat{e}_s) = 1 \text{ for } e = \hat{e}_s$$

$$= 0 \text{ elsewhere} \quad (12)$$

Then, by applying the compositional rule of inference [Equation (9)] and the centroid equation (11), it can be shown that

$$\hat{c}_q = \frac{\int c_q Proj[\mu_R(\mathbf{e},\mathbf{c})](\hat{e}_s,c_q)dc_q}{\int Proj[\mu_R(\mathbf{e},\mathbf{c})](\hat{e}_s,c_q)dc_q} \quad q=1,2,\cdots,p; \quad s=1,2,\cdots,n \quad (13)$$

This result is used in making control decisions in the case of a coupled rule base.

Inferencing through an Uncoupled Rule Base

Here, the rule base is assumed to be uncoupled, and subgroups of rules are considered separately, in which each subgroup relates just one condition variable e_s and one action variable c_q; thus

$$R_{s,q} : \text{ELSE[IF } e_s \text{ is } E_s^i \text{ THEN } c_q \text{ is } C_q^j] \quad (14)$$

Note that the membership function of the rule base subgroup is given by

$$\mu_{R_{s,q}}(e,c) = \max_{i,j(i)} \min[\mu_{E_s^i}(e), \mu_{C_q^j}(c)] \quad (15)$$

For a fuzzy context $\hat{E}_s$ that is given at a particular instant, the corresponding fuzzy action $\hat{C}_q$ is obtained by applying the compositional rule of inference in the usual manner, as

$$\mu_{\hat{C}_q}(c) = \sup_e \min[\mu_{\hat{E}_s}(e), \mu_{R_{s,q}}(e,c)] \quad (16)$$

If the context is assumed to be a fuzzy singleton of value e_s, the corresponding crisp action is determined by the centroid method; thus

$$\hat{c}_q = \frac{\int c \mu_{R_{s,q}}(\hat{e}_s,c)dc}{\int \mu_{R_{s,q}}(\hat{e}_s,c)dc} \quad (17)$$

Even though the computational requirements of inference making can be significantly reduced by the assumption of an uncoupled rule base, some accuracy is lost here. This is addressed in the next section. Note, however, that the concepts of a coupled rule base can be applied directly to an uncoupled rule base, through the use of the familiar concept of *cylindrical extension* (Zadeh, 1975), as described in Chapter 3, and defined below.

Definition

Consider $\mathbf{x} \triangleq [x_1, x_2, \ldots, x_n]^T \varepsilon \Re^n$, $\mathbf{y} \triangleq [y_1, y_2, \ldots, y_p]^T \varepsilon \Re^p$, and $\mu(x_i, \mathbf{y})$: $\Re \times \Re^p \to [0,1]$ with $1 \leq i \leq n$. The *cylindrical extension* of $\mu(x_i, \mathbf{y})$ over the entire space $\Re^n \times \Re^p$ is given by

$$Cyl[\mu(x_i, \mathbf{y})](\mathbf{x}, \mathbf{y}) = \mu(x_i, \mathbf{y}), \forall \mathbf{x}, \mathbf{y}; 1 \leq i \leq n$$

$$\Re^n \times \Re^p \to [0,1] \tag{18}$$

■

With this definition, it is clear that the membership function of the entire rule base is the *intersection* of the cylindrical extensions of the membership functions of the subgroups; thus

$$\mu_R(\mathbf{e}, \mathbf{c}) = \bigwedge_{s, q(s)} Cyl[\mu_{R_{s,q}}(e_s, c_q)](\mathbf{e}, \mathbf{c}) \tag{19}$$

The uncoupled rule base will not suffer any further loss of accuracy through this extension in view of the fact that in the uncoupled case we have

$$Proj\{Cyl[\mu_{R_{s,q}}(e_s, c_q)](\mathbf{e}, \mathbf{c})\}(e_s, c_q) = \mu_{R_{s,q}}(e_s, c_q) \tag{20}$$

Equivalence Condition

From the developments presented thus far, it is clear that the control inferences made using a coupled rule base are generally not the same as those made by assuming an uncoupled rule base for the same system parameters and data. However, it can be shown that if a certain condition is satisfied by the membership functions of the rule base variables, the control inferences in the two cases become identical, with an obvious computational advantage in the case of the uncoupled rule base. A condition for this equivalence is established in this section.

Zadeh (1975a, b) developed the concept of *separability of fuzzy restrictions*, and showed that a fuzzy restriction is separable if and only if the *join* of the projections of the restrictions results in the original restriction. Note that the join is the intersection of the cylindrical extensions. This concept can be extended to establish the equivalence condition mentioned above. Again, consider the coupled rule base given by Equation (8). Its membership function may be expressed as

$$\mu_R(\mathbf{e}, \mathbf{c}) = \max_{i,j(i)} \min[\mu_{E_1^{i(1)}}(e_1), \mu_{E_2^{i(2)}}(e_2), \cdots, \mu_{E_n^{i(n)}}(e_n),$$

$$\mu_{C_1^{j(1)}}(c_1), \mu_{C_2^{j(2)}}(c_2), \cdots, \mu_{C_p^{j(p)}}(c_p)] \quad (21)$$

where i denotes a coupled rule in the rule base, and hence represents the associated set of fuzzy states of the condition variables, and j denotes the associated set of fuzzy states of the action variables. The *min* operation of the membership functions in Equation (21) may be interpreted as the intersection of their cylindrical extensions in $\Re^n \times \Re^p$. Then, when using Equation (21) in Equation (13) for making inferences through a coupled rule base, it is required that the projection $\Re^n \times \Re^p \to \Re \times \Re$ given by

$$Proj[\mu_R(\mathbf{e}, \mathbf{c})](e_s, c_q) = \sup_{\substack{\forall e_k \neq e_s \\ \forall c_\ell \neq c_q}} \{\mu_R(\mathbf{e}, \mathbf{c})\} \quad (22)$$

be determined. Now, by substituting Equation (21) into (22) it can be shown that

$$Proj[\mu_R(\mathbf{e}, \mathbf{c})](e_s, c_q) \equiv \max_{i,j(i)} \min[\mu_{E_s^{i(s)}}(e_s), \mu_{C_q^{j(q)}}(c_q), \mu_i] \quad (23)$$

in which

$$\mu_i = \min_{\substack{k \neq s \\ \ell \neq q}} \sup_{e_k, c_\ell}[\mu_{E_k^{i(k)}}(e_k), \mu_{C_\ell^{j(\ell)}}(c_\ell)] \quad (24)$$

Then, by comparing Equation (23) with Equation (15) it follows that the equivalence condition is

$$\mu_i \geq \min[\sup_{e_s} \mu_{E_s^{i(s)}}(e_s), \mu_{C_q^{k(q)}}(c_q)] \quad (25)$$

A special case in which the condition (25) is satisfied is when all the fuzzy variables in the rule base are *normal* [i.e., when they have peak membership grades of unity or, *hgt* (membership function) = 1]. In this case $\mu_i = 1$ for all i. See Appendix A for a discussion of the function *hgt*.

■ EXPERIMENTAL ILLUSTRATION

To demonstrate the use of an uncoupled rule base in a realistic application, and to study the effect of uncoupled rules, the experimental DC

Context	Actions
RT = 95% Rise Time	PX = Phase Lead at Crossover
FD = Damped Natural Frequency	GX = Gain at Crossover
DR = Damping Ratio	XF = Crossover Frequency
OS = Response Overshoot	LF = Low Frequency of
OF = Response Offset	Integrator for Crossover Gain

Rule Base

1. For Rise Time:

   Else [If *RT* is *X* then *PX* is *A*]
   X, *A*(*X*)

   Else [If *RT* is *X* then *GX* is *A*]
   X, *A*(*X*)

   Else [If *RT* is *X* then *XF* is *A*]
   X, *A*(*X*)

2. For Damped Natural Frequency:

   Else [If *FD* is *X* then *GX* is *A*]
   X, *A*(*X*)

3. For Damping Ratio:

   Else [If *DR* is *X* then *PX* is *A*]
   X, *A*(*X*)

   Else [If *DR* is *X* then *GX* is *A*]
   X, *A*(*X*)

4. For Response Overshoot:

   Else [If *OS* is *X* then *PX* is *A*]
   X, *A*(*X*)

   Else [If *OS* is *X* then *XF* is *A*]
   X, *A*(*X*)

5. For Response Offset:

   Else [If *OF* is *X* then *GX* is *A*]
   X, *A*(*X*)

   Else [If *OF* is *X* then *LF* is *A*]
   X, *A*(*X*)

Figure 7 The rule base of servo tuning.

servomotor system of knowledge-based tuning that was described in the first part of this chapter is used. This application is consistent with the assertion that tuning is more appropriate than low-level direct control, as an application of fuzzy logic-based, on-line decision making.

The rule base that has been implemented in the servo motor system is uncoupled, as clear from Figure 7. First, the entire rule base is executed at high speed to carry out the tuning of an initially oscillatory system. The response of the servo motor with an inertial load, in this case, is shown in Figure 8(a). It could be argued that if reasoning is carried out sufficiently fast, characteristics of a coupled rule base would be retained here, at least to a limited extent. For example, the two uncoupled rules

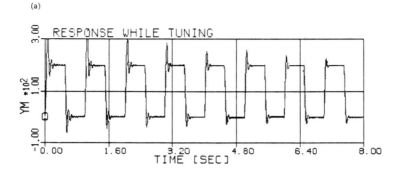

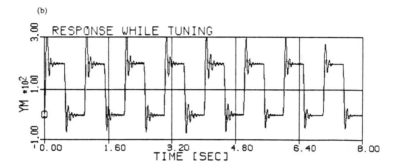

Figure 8 The response of the prototype servo motor during tuning (a) with all rules active, (b) with the rule subgroup sequence 1-2-3-4, (c) with the rule subgroup sequence 2-3-4-1, (d) with the rule subgroup sequence 3-4-1-2.

$$\text{IF } RT \text{ IS } X \text{ THEN } PX \text{ IS } A$$

and

$$\text{IF } FD \text{ IS } Y \text{ THEN } GX \text{ IS } B$$

may exhibit the behaviour of the coupled rule

$$\text{IF } RT \text{ IS } X \text{ AND } FD \text{ IS } Y \text{ THEN } PX \text{ IS } A \text{ AND } GX \text{ IS } B$$

if the uncoupled rules are executed very fast (i.e., almost in parallel), which is the case for the result shown in Figure 8(a).

Next, the four subgroups of rules, as denoted by 1 through 4 in Figure 7 were sequentially activated, for uniform intervals of time, one at a time. It was found that in each case proper tuning could be achieved if the sequence was cycled for a sufficient length of time. It follows that tuning accuracy is not compromised due to sequencing a set of uncoupled rules, instead of executing a coupled rule base. Because the membership functions in the rule base were chosen to be normal, this behavior was to be expected, in view of the

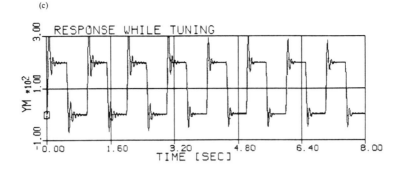

Figure 8 (continued).

equivalence result developed in this chapter. However, when the duration of a sequence was kept unchanged and the order in which the subgroups of rules are sequenced was changed, the end results were found to differ depending on the particular sequence. This too was to be expected because different subgroups of rules emphasize different characteristics of the servo response. Because these characteristics of the response vary with time, the end result of tuning depends on the time segment of the response in which a particular subgroup of rules is activated. Three such results are presented in Figures 8(b) through (d). In these experiments the switching sequence of the rule-base subgroups was maintained the same, in a cyclic sense, but the starting subgroups of rules were not identical. Hence, the efficiency of tuning is seen to depend on the time period during which a particular subgroup of rules is active.

■ SUMMARY

This chapter developed an intelligent tuner for a servo motor. A reference model was used, first to set performance specifications and then, during tuning, to determine the progress of the tuner.

The tuner consisted of a servo expert level at which the performance of the servo motor is evaluated and performance indicators are determined, and a knowledge base level at which tuning decisions are made using a fuzzy ruleset. The action variables of the knowledge base were a set of controller attributes which could be conveniently related to a set of performance specifications. A mapping module was employed to convert the controller attributes into tuning parameters. The system was implemented and evaluated on a commercially available servo motor system. The approach presented in this chapter is different from previous ones due to the use of a *reference model,* a *tuner mapping module,* and a *performance evaluator module.*

In a coupled rule base more than one condition variable may link with more than one action variable within the same rule. The second half of the chapter developed an analytical framework for decoupling such a rule base, with specific reference to fuzzy-logic control. The computational problem of fuzzy-logic control is greatly simplified through the use of an uncoupled rule base. Conditions were derived under which decision making through an uncoupled rule base becomes equivalent to that through a coupled rule base. The same laboratory prototype of single-axis servo motor system that was described in the beginning of the chapter which incorporated a knowledge-based tuner was employed to demonstrate the effective use of an uncoupled rule base that is equivalent to a coupled one.

■ PROBLEMS

1. Consider a fuzzy-logic based self-tuning controller for a DC servo system, as schematically shown in Figure 9. It is suggested that, instead of using this approach which employs a fuzzy rulebase, first a crisp *tuning curve* should be established, that uses a performance attribute of the control system as the independent variable. Then the tuning action may be established as a crisp decision, during operation of the system, simply by first computing the performance attribute using the system response, and then reading the tuning action from the pre-determined curve. Comment on this suggestion, emphasizing on its general feasibility.

2. Application of expert systems has been proposed for ship design. But, the development of the knowledge base is not straightforward. Knowledge may be acquired from published literature, experimental observation, and by interrogating domain experts. Then, one will have to address the "coupled"

nature of the knowledge base. In particular, address the following questions:
   (a) How could one determine which cause affects which performance and by how much? For example, motion sickness may result from various factors such as the frequency and amplitude of motion, type and quantity of food eaten and how long ago, and the age and level of health of the person.
   (b) Once an uncoupled rule base is generated, how could one assign priorities or weights for the rules, and resolve conflicts?

3. Consider a rule base of the form

$$R: \bigcup_{i=1,m} E^i \to C^i$$

where there are m fuzzy levels (fuzzy resolution) for the condition variable E. Show that as $m \to \infty$ for a fixed and finite support set for E, the rule base may be expressed as a crisp relation. Why is m cannot be very large in practice? How would an appropriate value be chosen for m?

4. In a fuzzy-logic based robot controller, it is attempted to improve the performance in the following manner:

By observing the change in performance due to a specific control action, the more effective rules are tagged and the associated improvement in performance is noted. Then, the rules are enhanced (e.g., by changing the membership functions; by adding a belief factor; by changing the scaling parameters) according to the required system behavior. Indicate difficulties in implementing such an approach.

5. Intuitively, fuzzy logic is appropriate for inherently fuzzy problems rather than crisp ones. But we come across many applications where the original problem is crisp; and is made fuzzy for the purpose of using fuzzy-logic techniques. Zadeh's *interpolative reasoning* is useful in these problems, as is the extension principle. In interpolative reasoning, a crisp relationship between two variables (say, u and v) may be replaced by a set of fuzzy points on the variable plane. Each point, then, is a two-dimensional membership function (in u and v). Discuss the use of interpolative reasoning in
   (a) an inherently fuzzy decision making problem
   (b) an originally crisp problem

Interpolative reasoning is widely used in fuzzy processor chips.

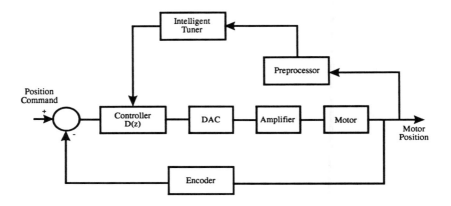

Figure 9 Block diagram of a fuzzy-logic based self-tuning system for a DC servo motor.

REFERENCES

de Silva, C.W. (1989). *Control Sensors and Actuators*, Prentice-Hall, Englewood Cliffs, NJ.

de Silva, C.W. (1991). An Analytical Framework for Knowledge-Based Tuning of Servo Controllers, *Eng. Appl. Artif. Intell.*, Vol. 4, No. 3, pp. 177–189.

de Silva, C.W. (1994). A Criterion for Knowledge Base Decoupling in Fuzzy-Logic Control Systems, *IEEE Trans. Syst. Man Cybernetics*, Vol. 24, No. 10, pp. 1548–1552.

de Silva, C.W. and MacFarlane, A.G.J. (1989). *Knowledge-Based Control with Application to Robots*, Springer-Verlag, Berlin.

de Silva, C.W. and Barlev, S. (1992). Hardware Implementation and Evaluation of a Knowledge-Based Tuner for a Servo System, *Proc. IFAC Symp. ACASP*, Grenoble, France, pp. 293–298.

Franklin, G.F., Powell, J.D., and Naeini, A.E. (1986). *Feedback Control of Dynamic Systems*, Addison-Wesley, Reading, MA.

Zadeh, L.A. (1973). Outline of a New Approach to the Analysis of Complex Systems and Decision Processes, *IEEE Trans. Syst. Man and Cybernetics*, Vol. SMC-3, No. 1, pp. 28–44.

Zadeh, L.A. (1975a). Calculus of Fuzzy Restrictions, in *Fuzzy Sets and Their Applications to Cognitive and Decision Processes*, L.A. Zadeh, K.S. Fu, K. Tanaka, and M. Shimura (Eds.), Academic Press, New York, pp. 1–39.

Zadeh, L.A. (1975b). The Concept of a Linguistic Variable and Its Application to Approximate Reasoning, Parts 1-3, *Inf. Sci.*, pp. 43–80, 199–249, 301–357.

Zadeh, L.A. (1979). A Theory of Approximate Reasoning, *Machine Intelligence*, Hayes, J. et al. (Eds.), Vol. 9, pp. 149–194.

Ziegler, J.G. and Nichols, N.G. (1942). Optimum Settings for Automatic Controllers, *Trans. ASME*, Vol. 64, pp. 759–768.

8 HIERARCHICAL FUZZY CONTROL

INTRODUCTION

Hierarchical control corresponds to an architecture that is commonly used in distributed control systems and is particularly relevant when various control functions in the system can be conveniently ranked into different levels (de Silva, 1989, 1990, 1991a,b; 1993a,b; 1995; de Silva and MacFarlane, 1988, 1989). The levels in a hierarchical control structure can be viewed from various points of view. The nature of the tasks carried out and information handled, the degree of resolution of information, and the bandwidth and event duration are considerations that can be used to rank various levels of a control hierarchy.

Example
An "iron butcher" is a head-cutting machine that is commonly used in the fish processing industry. Millions of dollars worth of salmon are wasted annually due to inaccurate head cutting using these somewhat outdated machines. The main cause of wastage is the "overfeed problem". This occurs when a salmon is inaccurately positioned with respect to the cutter blade so that the cutting location is beyond the collarbone and into the body of a salmon. Similarly, an "under-feed" problem occurs when the cutting is incomplete, and unpalatable parts of the fish head are retained with the processed fish body. An effort has been made to correct this situation by sensing the position of the collarbone using a vision system and automatically positioning the cutter blade accordingly (de Silva, 1990). A schematic representation of an electromechanical positioning system of a salmon-head cutter is shown in Figure 1. Positioning of the cutter is achieved through a lead screw and nut arrangement which

236 INTELLIGENT CONTROL: FUZZY LOGIC APPLICATIONS

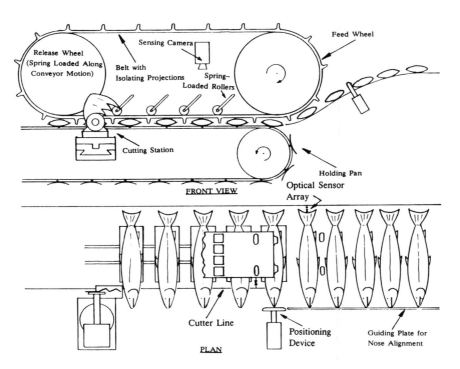

Figure 1 A positioning system for an automated fish cutting machine.

is driven by a brushless DC motor. The cutter carriage is integral with the nut of the lead screw and the AC motor that drives the cutter blade. The carriage slides along a lubricated guideway. In this system position control of the cutter and the speed control of the conveyor can be accomplished using conventional means (e.g., through servo control). However, suppose that in addition the quality of the processed fish is to be observed and on that basis several control adjustments are to be made. These adjustments may include descriptive (e.g., small, medium, large) changes in the cutter blade speed, conveyor speed, and the holding force. The required degree of adjustment is learned through experience. It is not generally possible to establish crisp control algorithms for such adjustments, and the control knowledge may be expressed as a set of rules such as (de Silva, 1993a):

> If the "quality" of the processed fish is "not acceptable", and if the cutting load appears to be "high", then "moderately" decrease the conveyor speed, or else if the hold-down force appears to be "low", then "slightly" increase the holding force, or else...

Such a set of rules forms the *knowledge base* of this control problem. It should be clear that this rule base has *qualitative, descriptive,* and *linguistic* terms such as "quality", "acceptable", "high", and "slight". These are fuzzy descriptors. Crisp representations and algorithms used in conventional control cannot directly incorporate such knowledge. Fuzzy logic control may be used in the high-level control of this machine.

This chapter addresses various considerations of knowledge-based hierarchical control with direct reference to an industrial application. The *degree of fuzziness* is viewed here as a means of classifying information and knowledge for various levels in the hierarchy. The concept of "fuzzy resolution", in particular, is addressed. This chapter also explores a hierarchical model for knowledge-based control of a process plant. The associated knowledge is assumed to be available in the form of linguistic statements containing fuzzy descriptors, and in view of this, the compositional rule of inference would be employed in the associated decision-making process (Zadeh, 1975, 1979). Information processing operations used in the model are examined, and their effect on the fuzziness of the information is studied (de Silva, 1993a, 1995).

▆ GENERAL CONCEPTS

Generally, the higher the hierarchical level, the more intelligent the control actions (de Silva, 1991a, 1993a). At higher levels, decisions may be made using incomplete information and yet the performance of the overall system may be acceptable. This is a characteristic of an intelligent system. Task description, knowledge representation, and decisions at these high levels of a hierarchy may be *qualitative* and *linguistic*, and hence, fuzzy, and consequently some intelligence will be needed for interpretation and processing of this information in order to make inferences (and control actions). Information that is *vague, ambiguous, general, imprecise,* and *uncertain,* even though not synonymous with fuzzy information, may be present at higher levels of control, and is also amenable to intelligent controllers. On the other hand, at low levels, a controller will typically need crisp, precise and nonfuzzy information (reference inputs, feedback signals, parameter values, etc.); otherwise, the system performance could deteriorate considerably.

Another distinction between various levels of a hierarchical system can be made in terms of the cycle time of the associated events and their control bandwidth. Due to the nature, amount, and complexity of the information that is needed at higher levels, the computational overheads will be higher and decision making will be slower as well. This is not a serious drawback because the cycle times of the events at higher levels of a hierarchical control system are generally long. In contrast, at low levels of hierarchy, faster action with short-term information will be needed as dictated by the process speed requirements, and as a result, the control bandwidth will be higher.

Information of high resolution can be intractable at high levels of a hierarchy due to the fact that for a specified amount of information that is required at a particular level, the amount of data needed will directly (and in many cases, exponentially) increase with the degree of resolution. Indeed, high resolution levels may not be needed at high hierarchical levels. Quite the opposite is generally true at low levels. This chapter formalizes several of these considerations.

Consider a process plant consisting of one or more workcells. Each workcell may consist of components such as robots, material transfer modules, processing devices, fixturing, and monitoring systems, with associated controllers, sensors, and actuators. The components must be properly interfaced or linked for automated and autonomous operation of the system. The performance of the system will depend strongly on the control and communication architecture that is employed. In particular, a control system consisting of a hierarchical architecture would be appropriate. Specifically, the system may be arranged into a functional hierarchy, with modules (both hardware and software) of similar functionality occupying a single layer (de Silva, 1991a; Isik and Meystel, 1986). Each module will possess a fixed input-output structure, and will represent an object in an *object-oriented paradigm*. At least some of the upper layers will contain knowledge systems for carrying out high-level control tasks. Because the higher layers of the architecture generally deal with low-resolution, imprecise, and incomplete information, more intelligence (or, knowledge-based action) would be needed in the associated decision-making process. Furthermore, high-resolution information from sensors in the lower levels must be properly "filtered" so as to provide compatible context for decision making through the upper-level knowledge systems.

Direct controllers, associated high-resolution sensors, and process actuators (de Silva, 1989, 1991a, 1993) may occupy the lowest level of the control system. Upper-level tasks may include process monitoring and tuning, reference command generation, component coordination, system reconfiguration, subtask assignment, task decomposition, and task planning in the ascending order. Such a hierarchical structure is shown in Figure 2. The end objective of the plant could be to cut salmon for various requirements (e.g., butchering for fresh and frozen markets, filleting, and portion control for

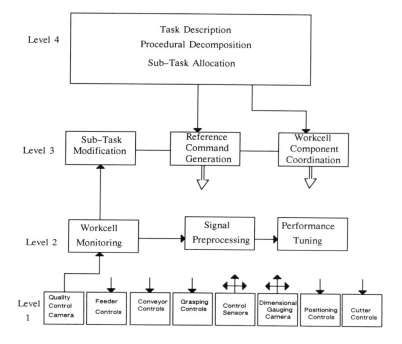

Figure 2 The functional hierarchy of a process workcell.

canning). In this context raw product recovery or yield, throughput, product quality, and aesthetic features become important subgoals. For a particular task such as head removal of a batch of salmon, the mechanical process involves feeding whole salmon into the machine (conveyor), grasping and holding down each salmon separately on the conveyor, sensing for the purpose of establishing the cutting locations, and cutting the salmon according to the sensory information. Various components of the workcell must be properly commanded and coordinated according to the requirements of the particular mechanical processing task. Furthermore, the overall system must be monitored at appropriate levels for performance evaluation and perhaps feedback control.

The hierarchical structure shown in Figure 2 has four levels. While the functions assigned to each level in this structure are not arbitrary, it is not difficult to increase the number of levels and enhance the functional distribution, thereby producing a more elaborate control structure. The present structure is adequate, however, for the purposes of this chapter. In particular, note the directed feedback paths from lower levels to upper levels. These exist in addition to the feedback paths in the same level and within the same functional module (block). First, consider the degree of resolution of the information at various levels of the hierarchy. At the highest level (level 4)

240 INTELLIGENT CONTROL: FUZZY LOGIC APPLICATIONS

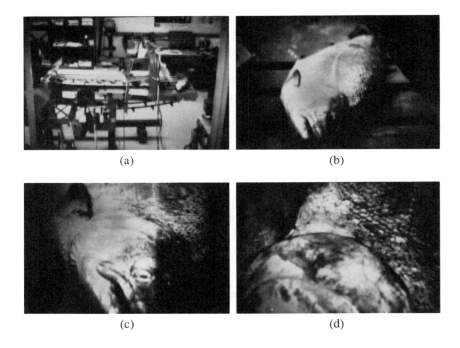

Figure 3 Information resolution at various levels of image in the same view.

a procedural description of a particular task is given, perhaps in a simple linguistic form. This description is then used to identify various subtasks which will be allocated to appropriate components of the workcell. At this level the required degree of information resolution is relatively low. As one proceeds down the hierarchy, the information resolution must increase, and in particular, the best available resolution should be used in low-level direct control (e.g., servo control) of the workcell components. To explain this further in the context of image data, consider Figure 3.

A view of a fish processing machine in our laboratory, as captured by a CCD camera of effective pixel resolution 510 × 492, is shown in Figure 3(a). By approaching the machine along the line of view of the camera, the image shown in Figure 3(b) is obtained. Further zooming-in along the same direction produced the subsequent two images given by Figures 3(c) and (d). It is now clear that the line of view has been directed at a fish on the conveyor, but this fact was not clear from Figure 3(a). The information resolution has increased in the successive images. If the positioning of the cutter is based on accurate gauging of the gill (or collarbone) position, it is clear that Figure 3(d), and not Figure 3(a), is the image that contains the necessary information. On the other hand, the information in Figure 3(a) can be quite useful in high-level monitoring of the workcell. Intermediate degrees of resolution will be useful in monitoring tasks at intermediate levels, for example, in establishing the product quality. To increase the resolution of Figure 3(a) to equal that of Figure 3(d), much more data would be needed; however, if the objective is to detect the gill position accurately, it is not necessary to increase the resolution throughout Figure 3(a) because most of the added data would not be useful for the purpose. Alternatively, the information in Figure 3(a) at the existing low resolution will be quite adequate for purposes such as:

1. Detecting the presence/absence of a fish.
2. Triggering and coordinating the components (e.g., feeder, vision system, cutter) of the workcell.
3. High-level monitoring/troubleshooting of the system (e.g., jamming, safety hazards).

A quality control decision of "Accept" or "Reject" may depend on a variety of factors, including information of high resolution obtained through low-level image processing. Then the high-level decision itself has a low fuzzy resolution, while the low-level information used to arrive at such decisions has a high resolution. As noted before, information (sensing, control, processing) bandwidth and event duration are also related to the level of control hierarchy. Specifically, the bandwidth decreases and the event duration increases with the level of hierarchy. Figure 4 summarizes the qualitative relationships of information resolution, control bandwidth, and fuzziness with the level of control hierarchy.

■ HIERARCHICAL MODEL

Procedural descriptions of tasks and subtasks at the top level of the hierarchy shown in Figure 2 may be done in a linguistic form. For example, statements such as "high-grade filleting for export" may contain fuzzy, vague, ambiguous, general, uncertain and imprecise terms, and interpretations. In particular, a high level of fuzziness may be tolerated at high levels of the control hierarchy, while even a slight fuzziness at the lowest level can lead to inaccurate control. In this context the meanings of "fuzzy resolution" and "degree of fuzziness" are important.

Fuzzy Resolution
The total number of possible fuzzy states of a variable within a given knowledge base (universe) represents the fuzzy resolution of that

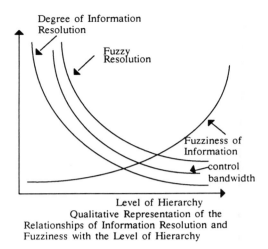

Figure 4 Qualitative representation of the relationships of information resolution and fuzziness with the level of hierarchy.

variable (de Silva, 1991b). For instance, suppose that the knowledge base at some level of the control hierarchy is expressed by the set of linguistic rules:

$$\underset{i,j,\cdots,k}{\text{ELSE}}[\text{IF } E_1^i \text{ THEN IF } E_2^j \text{ THEN IF } \cdots E_n^k \quad \text{THEN } C_\ell^a \text{ AND } C_q^b \text{ AND } \cdots C_p^c] \quad (1)$$

where the condition variable (antecedent) e_j may assume m discrete fuzzy states $E_j^1, E_j^2, \ldots, E_j^m$ for j 1,2, ..., n, and similarly each action variable (consequent) c_j may assume r discrete fuzzy states $C_j^1, C_j^2, \ldots, C_j^r$ for $j = 1,2, \ldots, p$. Note that the integer value m represents the fuzzy resolution of E_j, and similarly, r represents the fuzzy resolution of C_j. The fuzzy resolution of a variable may be interpreted through the nature of the membership of that variable as well. Specifically, each fuzzy state of the variable will have a corresponding modal point in the membership function (typically, having a peak membership grade). Then, the number of such modal points in the membership function (within its support set) corresponds to the number of fuzzy states of the variable (within the knowledge base). Accordingly, the *intermodal spacing* of the membership function may be used as a measure of the fuzzy resolution of the variable. This interpretation is illustrated in Figure 5. Then, an appropriate quantitative definition for fuzzy resolution r_f would be

$$r_f = \frac{w_s}{w_m} \quad (2)$$

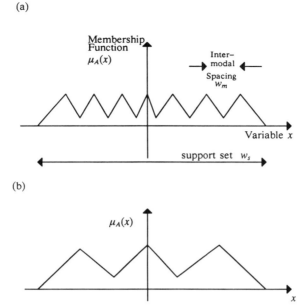

Figure 5 An interpretation of fuzzy resolution. (a) High fuzzy resolution. (b) Low fuzzy resolution.

in which
w_s = width of the support set of A
w_m = intermodal spacing
for a membership function

$$\mu_A(x): \Re \to [0,1] \tag{3}$$

of a fuzzy set A. Note that as the fuzzy resolution increases, the information resolution also increases, and the corresponding knowledge base becomes more appropriate for a lower level of the control hierarchy. When the membership function shape of a fuzzy state is trapezoidal rather than triangular, the centroidal locations may be considered as the modal points in the present definition of fuzzy resolution.

Degree of Fuzziness

The degree of fuzziness of a fuzzy set is a measure of the difficulty of ascertaining the membership in the fuzzy set of the elements of the support set. It follows that if the membership grade of an element is <0.5, it is easy to exclude that element from the set, and if it is >0.5 it is easy to include the element. Accordingly, the most fuzzy elements are those whose membership grade is 0.5. Many quantitative measures are available (Dubois and Prade, 1980). One measure of the degree of fuzziness of a fuzzy set (A) may be defined with respect to a fuzzy state or a modal point in A, and expressed as

$$d_f = \frac{1}{m} \int_X f(x)dx \qquad (4)$$

in which the non-negative function $f(x)$ is defined in the support set X of the fuzzy set A by

$$f(x): X \rightarrow [0, 0.5] \qquad (5)$$

with

$$\begin{aligned} f(x) &= \mu_A(x) \quad \text{for} \quad \mu_A(x) \leq 0.5 \\ &= 1 - \mu_A(x) \text{ otherwise} \end{aligned} \qquad (6)$$

and m is the number of modal points in A. This definition assumes continuous (piecewise) membership function, but may be extended to the discrete case as well. A graphical representation of this measure of fuzziness is given in Figure 6(a). The present definition is given on a "per mode" basis.

Another representation of fuzziness may be given, as indicated in Chapter 3. If the membership grade of an element is close to unity, the element is almost definitely a member of the set. Conversely, if the membership grade of an element is close to zero, the element is nearly outside the set. Accordingly, the fuzziness of a set may be measured by the closeness of its elements to the membership grade of 0.5. In view of this, a measure of fuzziness of set A would be

$$f_A = \int_{x \in X} \left| \mu_A(x) - \mu_{A_{1/2}}(x) \right| dx \qquad (7)$$

Here, $\mu_{A_{1/2}}$ is the $1/2$-cut of A, which is defined next, as in Chapter 3.

α-Cut of a Fuzzy Set

The α-cut of a fuzzy set A is the crisp set formed by elements A whose membership grades are greater than or equal to a given value α. This set is denoted by A_α. Specifically,

$$A_\alpha = \{x \in X | \mu_A(x) \geq \alpha\} \qquad (8)$$

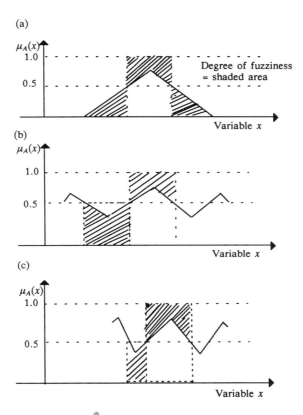

Figure 6 The degree of fuzziness: (a) definition, (b) high degree, and (c) low degree.

in which X is the universe to which A belongs. Further details are provided by Dubois and Prade (1980) and de Silva (1993a).

Analytical Model

Now let us return to a hierarchical control structure of the type given in Figure 2. Our emphasis here is on the knowledge-based control aspects of the system, including controller tuning and decision making at higher levels. First, it is assumed that the overall hierarchical system has been designed using conventional techniques. This includes task allocation, resource allocation, system optimization, and low-level control. Our approach, then, is to develop a supplemental hierarchical system that uses additional knowledge and experience to further improve the performance of the system. In particular, knowledge that may be expressed in a linguistic and qualitative form and which is amenable to fuzzy-logic processing is considered.

An analytical model for a fuzzy-knowledge-based, hierarchical control system is shown in Figure 7. What is shown is a general layer

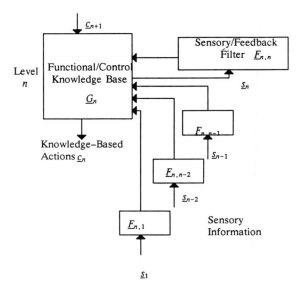

Figure 7 A supplemental knowledge-based hierarchical system.

(level n) of the knowledge-based structure. This particular structure of information transfer and control can have a variety of applications. Functional modules are shown in the figure as blocks, and the signal and communication paths are indicated as directed lines. In particular, we observe that there are "action lines", corresponding to inferences, commands, and dynamic reference levels that go from an upper level to a lower level, and "sensing lines", which are lateral toward the sensory/feedback blocks that produce feedback signals.

The "feedback lines" can occur both at the same level and from a lower level to an upper level. In the latter case, a filter/interpreter is used to convert the information into a form that is consistent with the data acquisition capabilities of the upper level. The notation **G** is used to denote the functional and control knowledge base at a particular level of the hierarchy, and it contains structured information in the form of linguistic fuzzy rules. The knowledge base receives instructions and inferences **c** from the upper level and also information and specialized knowledge **s** from sensory feedback and knowledge sources at the present level and the lower level. When sensory information is received from the lower level, an information filter, preprocessor, or an interpreter (denoted by **F**) will be necessary in order to generate the corresponding feedback commands. This hierarchical structure is operationally expressed as (de Silva, 1991a, 1993a)

$$\mathbf{c}_n = \mathbf{G}_n \otimes [(\mathbf{F}_{n,n} \otimes \mathbf{s}_n) \oplus (\mathbf{F}_{n,n-1} \otimes \mathbf{s}_{n-1}) \oplus \cdots \oplus (\mathbf{F}_{n,1} \otimes \mathbf{s}_1) \oplus \mathbf{c}_{n+1}] \text{ for level } n \quad (9)$$

in which
- $\mathbf{F}_{i,j}$ = information prefilter from level j to level i
- $\mathbf{s}_j$ = information vector from level j
- $\mathbf{G}_i$ = knowledge base at level i
- $\mathbf{c}_i$ = decision vector at level i
- $\oplus$ = a combinational operator
- $\otimes$ = a transitional operator

The knowledge bases $\mathbf{G}_i$ and $\mathbf{F}_i$ may take linguistic, fuzzy-logic representations. Also, the operations $\oplus$ and $\otimes$, which are subjective and have interpretations depending on the specific application, may be based on various fuzzy logic operations and the compositional rule of inference. In particular, these operations can have a direct influence on the fuzziness of various information components that are being preprocessed and combined.

Because intelligence may be considered the capacity to acquire and meaningfully apply knowledge within the context of perception, reasoning, learning, and inference from incomplete knowledge, it is clear that the concept is directly related to the interpretation of the transitional operator $\otimes$ and the combinational operator $\oplus$. In the context of fuzzy inference, then, these operators can be interpreted within the framework of the compositional rule of inference, for example, using the *max-min* composition.

Feedback/Filter Modules

In Figure 7 the feedback filter (preprocessor) modules are denoted by $\mathbf{F}$. The nature of information processing that is required in each such module will depend on several factors such as the level and the level differential which the $\mathbf{F}$ module serves. In particular, when the level is high and the level differential is high, then high-level information processing will be involved in general. The cycle times of processing will be longer and, furthermore, information interpretation will become more prominent than direct processing. Formally, the transitional operation of a feedback filter, which is given by

$$\bar{\mathbf{s}} = \mathbf{F} \otimes \mathbf{s} \tag{10}$$

may mean

$$\bar{\mathbf{s}}(\bar{t}) = \bar{\mathbf{s}}(\mathbf{s}[t,T]) \tag{11}$$

with

$$\mathbf{s} = [s_1, s_2, \cdots, s_n]^T \tag{12}$$

$$\bar{\mathbf{s}} = [\bar{s}_1, \bar{s}_2, \cdots, \bar{s}_m]^T \tag{13}$$

and $m \leq n$. Here, T denotes the cycle time of the operation which represents the duration of the input information **s** that is needed to generate one instant of inferences (outputs) $\bar{\mathbf{s}}$ within the **F** module. Note that $\bar{\mathbf{s}}$ represents part of the context that is needed by the functional control module **G** at the particular layer. The number of inference channels m of **F** is generally less than the number of input information channels n in high-level processing.

Example
Inputs to a feedback/filter module for level 2 of the fish processing workcell of Figure 2 may be cutter blade speed, conveyor speed, and images of processed fish over a duration of 60 s. The output inference made by **F** might be a single "quality index" of processing for over 100 fish that are processed during this period. Peak variations, average values, and shape factors may be involved in establishing the quality index "value". The value itself may be of the type "acceptable/unacceptable".

Because the membership grades of the input information to **F** are processed through *min*-type norms corresponding to AND or cross-product operators, it should be clear that the inferences of **F** generally will be fuzzier than the input information. To further illustrate this, consider a "data averaging" function. Then, assuming that the average "mapping" itself is crisp and the data set alone is fuzzy, we may represent the transitional operator $\otimes$ by

$$\bar{s}_j(\bar{t}) = \frac{1}{T} \int_{\tau}^{\bar{\tau}+T} s_j(\tau) d\tau \qquad (14)$$

Then, the membership grade of the filter output is given by

$$\mu_{\bar{s}_{j(\tau)}} = \min_{[\tau, \bar{\tau}+T]} [\mu_{s_{j(\tau)}}] \qquad (15)$$

This aspect of information preprocessing is further discussed later in the chapter.

Functional/Control Modules
In Figure 7 the functional and control modules are denoted by **G**. In the present model each module **G** contains a linguistic knowledge base, and makes decisions (actions) **c** using the context as determined by the decisions of the upper level as well as inferences made by the related feedback/filter modules. Then, Equation (9) may be rewritten as

$$\mathbf{c}_n = \mathbf{G}_n \otimes [\bar{\mathbf{s}}_n \oplus \bar{\mathbf{s}}_{n-1} \oplus \cdots \oplus \bar{\mathbf{s}}_1 \oplus \mathbf{c}_{n+1}] \tag{16}$$

Here, the combinational operator will take appropriate interpretations depending on the context information that is needed by the functional control module and the inferences that are provided by feedback/filter modules. For example, it may mean a *sup* or *max* operation for membership functions of like elements in various vectors on the right-hand side of Equation (16). Generally, then, the control decisions and actions **c** are established by applying the compositional rule of inference. Such an application to tuning the joint controllers of a robot is discussed by de Silva and MacFarlane (1988) and was presented in Chapter 6. The combinational operation is discussed further in the next section.

The upper level functional/control modules **G** will have learning and self-organization capabilities as well. For example, the membership functions corresponding to decisions/actions that apparently produce a worsening performance may be tagged. Then, the tagging statistics may be used to modify those membership functions and even to delete the corresponding linguistic rules altogether from the knowledge base. Similarly, the inferences that tend to improve the system performance may be tagged as well and subsequently used to enhance the corresponding rules and membership functions. Furthermore, new rules may be added to the knowledge base on the basis of experience and new knowledge. The functional/control module **G** need not be complete for the purpose of the supplemental knowledge system. At higher levels, it should be able to operate at a high level of intelligence with fuzzier, incomplete, and approximate information of relatively low resolution.

■ EFFECT OF INFORMATION PROCESSING

This section explores the effect of preprocessing (transitional operation $\otimes$) on the fuzziness of the result. This has direct implications in hierarchical control systems, as low-level information is preprocessed for use in high-level decision making, as discussed in connection with the model given in Figure 7.

Suppose that the membership function μ_c of the control action C is triangular and symmetric, with a unity modal value and a support set of $[-d,d]$, as shown in Figure 8. Also, suppose that a set of low-level signals is measured and fuzzified to give membership grades m_i and that

$$m = \min_i(m_i) \tag{17}$$

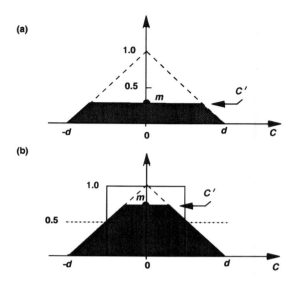

Figure 8 Control inference and its $1/2$-cut when (a) the context membership is <0.5 and when (b) the context membership is ≥0.5.

If the knowledge base uses the fuzzy quantities corresponding to m_i as the context, then the control inference for these measurements is obtained by applying the compositional rule, as given in Chapters 3 and 4. Clearly, the shaded area in Figure 8 corresponds to the control inference C'.

To obtain the fuzziness of C', according to Equation (7), we need to determine the $1/2$-cut of C'. Two cases exist: (a) $m < 0.5$, as shown in Figure 8(a), and (b) $m \geq 0.5$, as shown in Figure 8(b).

It should be clear in case (b) that $\mu_{C'_{1/2}}$ is the rectangular pulse shown in Figure 8(b). Then, using Equation (7), the following result is obtained:

$$f_{C'} = dm(2 - m) \quad \text{for} \quad m < 0.5$$
$$= \frac{d}{2}(3 - 4m + 2m^2) \quad \text{for} \quad m \geq 0.5 \qquad (18)$$

The relationship given by Equation (18) is plotted in Figure 9. It is seen from this figure that the fuzziness of the control inference peaks at $m = 0.5$, which is the case when the combined membership grade of the context variables is 0.5. In other words, the control inference is most fuzzy when the fuzzy state of the dominant sensor reading is also most fuzzy ($\mu = 0.5$).

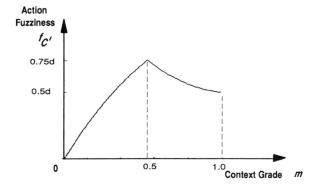

Figure 9 Variation of the fuzziness of the control inference with the membership grade of the rule base context.

Effect of Signal Combination on Fuzziness

Next, consider decision making with multiple rules and multiple contexts in which the combinational operation $\oplus$ would be applicable. The fuzziness of the control inference is given by

$$f_{C'} = \int_{c \in Z} \left| \mu_{C'}(c) - \mu_{C'_{1/2}}(c) \right| dc \tag{19}$$

This can be written as

$$f_{C'} = \int_{c \in z} \left| \max_i [\mu_{C'_i}(c)] - \max_j [\mu_{C'_{j_{1/2}}}(c)] \right| dc \tag{20}$$

Note that Z is the universe of discourse (or at least the support set) of C. Equation (20) follows from the fact that

$$\alpha\text{-cut}(A \cup B) = (\alpha\text{-cut } A) \cup (\alpha\text{-cut } B) \tag{21}$$

which may be easily verified. Next, in view of

$$\left| \max_i x_i - \max_j y_j \right| \le \max_i \left| x_i - y_i \right| \tag{22}$$

we have

$$f_{C'} \le \int_{\alpha \in Z} \max_i \mu_{C'_i}^f \, dc \tag{23}$$

Here, $\mu^f_{C'_i}$ is the local fuzziness at value c of the control inference C'_i, as given by

$$\mu^f_{C'_i}(c) = \left| \mu_{C'_i}(c) - \mu_{C'_{i_{1/2}}}(c) \right| \qquad (24)$$

Hence, Equation (23) states that under the combinational operation, the fuzziness of the result is bounded by the envelope of the local fuzziness curves of the individual components of information.

APPLICATION IN PROCESS CONTROL

An application of a knowledge-based hierarchical system is described in this section (de Silva and Wickramarachchi, 1992; de Silva, 1990, 1995). Fish processing is rapidly changing from a resource-driven industry to a market-oriented one. The reasons for this include the decline of fish stocks in general, high cost of processing — particularly the labor cost, and the high cost of fish itself. In Canada salmon is no longer intended for the fresh, frozen, and canning markets alone, and fish processors are increasingly concentrating on value-added products such as salmon steaks and roasts. Manual processing of fish is slow, inefficient, hazardous, tiring, and generally unpleasant. The high cost of maintenance, modification and replacement of existing machinery, and the lack of flexibility in terms of reconfiguration and upgrading have caused serious difficulties within the industry. These machines are known to be outdated (e.g., pre-World War II vintage), inaccurate, and wasteful. In view of these considerations, the fish processing industry must make a concerted effort to steadily move toward utilizing modern technology. To facilitate this, particularly for the salmon-processing industry in Canada, a laboratory has been established at the University of British Columbia along with the inauguration of an industrial research chair (de Silva, 1990). The general goal of the laboratory is to develop technology for the flexible automation of fish processing. Specifically, a laboratory workcell is being developed which incorporates computer vision, knowledge-based control, and customized mechanical designs, including robotic handling and processing devices. These developments are in the pre-competitive stage at present. Initially, the workcell is intended for accurately and efficiently removing the head of salmon. Even though high throughput rates are desirable, in the present market-driven industry the product quality is equally important. In view of this, incorporating local controllers at the bottom level of the workcell, a hierarchical control system is being developed for the fish processing system. This section specifically considers an intelligent monitoring system that has been developed for improving the "quality" of processed salmon while maintaining a sufficient throughput rate. The section describes the hardware and software details of the hierarchical structure, system operation, underlying principles and concepts, and some typical results.

Workcell Development

A control and communication diagram of the laboratory workcell is shown in Figure 10. This shows the main components and the low-level control structure of the system. A view of the laboratory system is shown in Figure 11. The fish fed at one end of the conveyor are continuously imaged by a CCD camera equipped with an electronic shutter. The vision workstation captures and analyzes each image to establish the proper cutting locations that are used in the lateral positioning of the cutter (Riahi and de Silva, 1990). A separate, ultrasound displacement sensor is used to measure the thickness of the fish near the gill area. This measurement is also needed in determining the lateral positioning of the cutter, if the direction of view of the camera is oblique. Furthermore, this measurement is needed to properly position the fish in the vertical direction prior to feeding into the cutter. Two DC servo motors, controlled by two drive axes of a position controller, are used to laterally position the cutter and vertically position the fish with respect to the cutter. Two blades arranged in a V-shape and driven by two induction motors through flexible shafts form the cutter assembly. This arrangement, along with the lateral positioning of the cutter and the vertical positioning of the fish, is intended to minimize the wastage of salmon during the cutting operation. There is a flat-belt hold-down mechanism for the entire batch of fish on the conveyor in order to prevent the movement of fish with respect to the conveyor, from the imaging location up to the cutting station. Also, there is a secondary, active, hold-down mechanism to hold the fish properly during cutting. The typical operating throughput is two fish per second. As a result, the low-level positioning controllers are required to have a cycle time on the order of 100 ms, with a total cutting time of approximately 300 to 400 ms and an image-capture, processing, and data communication time not exceeding 500 ms.

Hierarchical System

The fish processing workcell shown in Figure 10 has many components which must be operated and properly coordinated in order to perform its processing task. The overall operation will involve different types of actions such as sensing, direct control, high-level image processing, performance monitoring, component coordination, command generation, subtask modification, task description, procedural decomposition, and subtask allocation (de Silva, 1991a). Such actions may be ranked into several levels of functions and accordingly

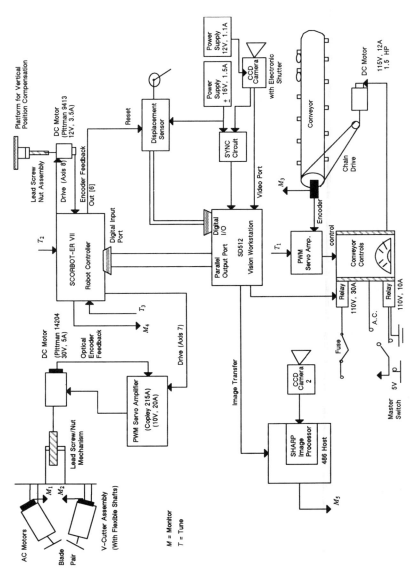

Figure 10 Control and communication diagram of the fish processing workcell.

HIERARCHICAL FUZZY CONTROL 255

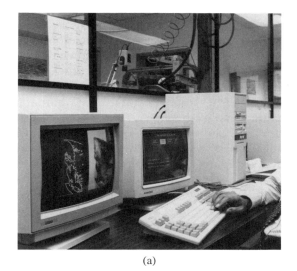

(a)

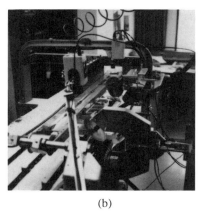

(b)

Figure 11 A view of the fish processing workcell. (a) Computer screen. (b) Workcell.

organized into a hierarchical control system (Isik and Meystel, 1986; Saridis, 1988; Albus et al., 1982; de Silva and MacFarlane, 1988). Even though a multilevel system encompassing a variety of actions within the workcell is being developed, only a three-level substructure intended for monitoring the workcell for cutting quality and making intelligent tuning decisions on that basis, is considered in this section. The lowest layer contains various components of the workcell such as sensors, actuators, and their direct controllers. The local control takes place at this level. Typically, the reference conditions for these controllers are preset and not altered at this level. It is expected that the operating conditions will vary, however, and thus the performance of the system components. The second layer of the hierarchy monitors various sensory signals from level 1 and preprocesses them. The preprocessed information is communicated to level 3 of the system, which contains a knowledge-based decision maker for system adjustment (tuning). The knowledge-based system makes tuning decisions by using the preprocessed sensory information as the context. The variables monitored are denoted as M_i and the possible tuning adjustments as T_j in Figure 10. The associated communication interface with the monitoring and intelligent tuning system and the internal software structure of this system are schematically shown in Figure 12. The software modules are programmed in C, and themselves are arranged in a functional hierarchy. The monitored variables are read through the system input/output board into the *sensory inputs module*. They undergo high-level processing in the *signal processor*, the results are passed to the *workcell status module*, and form the context for the knowledge-based system. The *knowledge base module* contains the rule base governing the intelligent tuning. The *decision maker module* computes the inferences corresponding to the context, using the knowledge base. The decisions are converted into physical adjustments in engineering units in the *tuning and control actions module*, and are communicated to the appropriate controllers of the workcell components through the system input/output board. The monitored variables include:

1. Speed of the upper blade of the cutter (M_1)
2. Speed of the lower blade of the cutter (M_2)
3. Conveyor speed (M_3)
4. Cutter position response (M_4)
5. Response of the vertical positioning platform (M_5)
6. Visual information of the processed fish (M_6)

The parameters adjusted by the control system include:

1. Speed setting of the conveyor (T_1)
2. Damping of the conveyor motion (T_2)
3. Proportional-integral-derivative (PID) parameters of the cutter positioning servo (T_3, T_4, T_5)
4. PID parameters of the vertical positioning platform servo (T_6, T_7, T_8)
5. Hold-down control (T_9)

The speed profiles of the cutter blades are first converted into cutter load profiles and expressed as a percentage of the full load of each cutter motor. These load profiles are then used to establish several attributes of cutting such as peak cutting load (CTLD), symmetry of cut (CTSM), and slipping of fish in the hold-down device (SLIP). The conveyor speed signal is checked to see whether the upper and lower speed limits (CNSL) are satisfied, and whether the speed fluctuations (CNOS) are significant. The position response of the cutter is used to determine attributes such as oscillatory or unstable response (CTOS), speed of response (CTSP), and steady-state error or

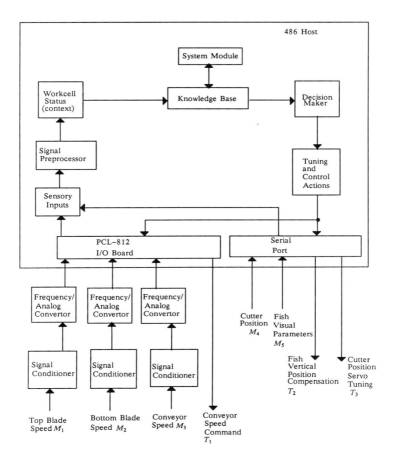

Figure 12 The hierarchical architecture of the workcell-monitoring and intelligent-tuning system.

offset (CTOF). Similarly, the response of the positioning platform is used to determine oscillatory or unstable response (PLOS), speed of response (PLSP), and offset (PLOF). Visual information obtained through image processing can provide supporting evidence for the symmetry of cut (CTSM), poor holding (SLIP), and excessive cutting load (CTLD), in addition to the cutting accuracy (CTAC). Specifically, the smoothness of the cut, continuity of the cut profile, and the depth of the cut may be obtained by high level processing of an image of a processed fish. The associated details are not presented here for the sake of brevity. Attributes obtained in this manner are the condition variables of the rule base: CTLD, CTSM, SLIP, CNSL, CNOS, CTOS, CTSP, CTOF, PLOS, PLSP, PLOF, and CTAC. The action variables are T_1: CNSP, T_2: CNDP, T_3: CTPG, T_4: CTIT, T_5: CTDT, T_6: PLPG, T_7: PLIT, T_8: PLDT, and T_9: HLDN.

Context			Actions	
CTLD =	cutting load	CNSP =	conveyor speed
SLIP =	slipping at holder	CNDP =	conveyor response damping
CTSM =	symmetry of cut	HLDN =	hold–down control
CTAC =	accuracy of cut	CTPG =	proportional gain of cutter
CNSL =	conveyor speed (limits)	CTIT =	integral time constant of cutter
CNOS =	conveyor oscillations	CTDT =	derivative time constant of cutter
CTOF =	cutter position offset error	PLPG =	proportional gain of platform
CTOS =	cutter oscillations		
CTSP =	cutter speed of response	PLIT =	integral time constant of platform
PLOF =	platform offset error	PLDT =	derivative time constant of platform
PLOS =	platform oscillations		
PLSP =	platform speed of response		

Rule Base

Else [If CTLD is X then CNSP is A]
X
$A(X)$

Else [If SLIP is X then CNSP is A and HLDN is B]
X
$A(X), B(X)$

Else [If CTSM is X and PLOF is Y then PLPG is A and PLIT is B]
X,Y
$A(X,Y), B(X,Y)$

Else [If CTAC is X and CTOF is Y then CTPG is A and CTIT is B]
X,Y
$A(X,Y) B(X,Y)$

Else [If CNSL is X then CNSP is A]
$X, A(X)$

Else [If CNOS is X then CNSP is A and CNDP is B]
X
$A(X), B(X)$

Else [If CTOS is X then CTPG is A and CTIT is B and CTDT is C]
X
$A(X), B(X), C(X)$

Else [If CTSP is X then CTPG is A and CTDT is B]
X
$A(X), B(X)$

Else [If PLOS is X then PLPG is A and PLIT is B and PLDT is C]
X
$A(X), B(X), C(X)$

Else [If PLSP is X then PLPG is A and PLDT is B]
X
$A(X), B(X)$

Figure 13 The knowledge base of the top-level decision maker.

The knowledge base that has been implemented in the fish processing workcell is summarized in Figure 13. Note that X and Y denote fuzzy variables that correspond to the condition variables of the rules which can take different fuzzy states (such as LOW, HIGH, VERY HIGH). The fuzzy variables A, B, and C denote the action variables and will assume the appropriate fuzzy values depending on the particular context. It follows that each rule given in Figure 13 is actually a composite of subrules combined through an *ELSE* connective. The notation in Figure 13 is used for the sake of brevity.

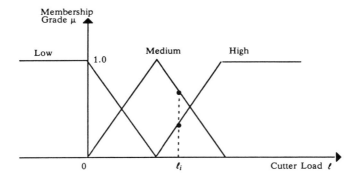

Figure 14 Membership functions for a CTLD description.

Information Prefilters

Interpretation of the monitored information must be carried out subjectively by taking into account the knowledge base requirements. Due to the limitation of space, only the cutter load variable CTLD is discussed here. The load of each cutter is defined as a percentage of the full load of the induction motors that are used to drive the cutter blades:

$$\text{Cutter Load} = \frac{\text{Slip}}{\text{Slip at Full Load}} \times 100\% \qquad (25)$$

where Slip is given by the difference of the synchronous speed and the actual speed.

For cutting quality and overload considerations it is the peak values of cutter load that are of primary interest. Initially, a simple and straightforward approach is taken in defining the fuzzy variable CTLD. Here, a three-level resolution of low, medium, and high is used and the associated membership functions are chosen as in Figure 14. Then, for a peak cutter load of ℓ_i the corresponding membership grades can be determined for the three categories of fuzzy description: low, medium, and high. This will form the fuzzy context, which will be directly used in the knowledge base for making inferences. The situation becomes more complex when there are several significant peak loads ℓ_1, ℓ_2, ... in a given cutter load signal. The simplest approach, then, would be to use the largest peak. Alternatively, some form of averaging may be applied for the membership grades of the various peaks, but the actual time locations of the peaks may also have some significance in terms of the cutting

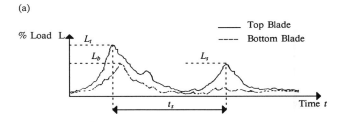
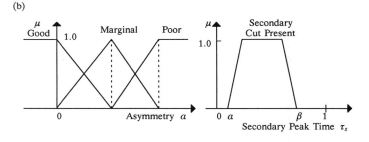
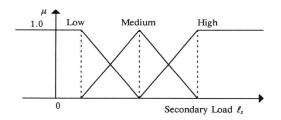

Figure 15 (a) Parameters of a cutter load profile. (b) Membership functions for asymmetry, secondary peak time, and secondary load.

quality. The interpretation of CTLD must be properly enhanced to take such considerations into account.

Consider a load profile of the cutter, as shown in Figure 15(a). The load L is expressed as a percentage, as given in Equation (25). Using a peak-detection algorithm, the primary peak loads L_t and L_b of the top and bottom blades of the cutter and possible secondary peak loads L_s for either of the cutter blades are determined. Also, the time t_s from the primary peak to such secondary peaks is determined. From these values, the following nondimensionalized indices are computed:

$$\text{Peak-load } L_p = max(L_t, L_b) \tag{26}$$

$$\text{Asymmetry index } a = \frac{|L_t - L_b|}{(L_t + L_b)/2} \qquad (27)$$

$$\text{Secondary peak - load index } \ell_s = \frac{L_s}{(L_t + L_b)/2} \qquad (28)$$

$$\text{Secondary peak - time - interval index } \tau_s = \frac{t_s}{T} \qquad (29)$$

in which T is the period between cuts (i.e., $1/T$ is the feed rate of fish).

The membership functions shown in Figure 15(b) are used to fuzzify these indices. The membership functions used for a and ℓ_s are similar, but not identical. For a computed value of an index, the grade of membership corresponding to each fuzzy state is determined from these membership function curves. Note that the asymmetry index is directly present as a context (CTSM) in the rule base. The other two indices (ℓ_s and τ_s) must be combined to provide the "fish slipping at the holder" (SLIP) context for the rule base. Specifically, the following rule base module is used:

IF Secondary Peak (τ_s) IS Present AND Secondary Load (ℓ_s) is Medium THEN

SLIP is Moderate :

ELSE IF Secondary Peak (τ_s) is Present AND Secondary Load (ℓ_s) is High (30)

THEN SLIP is Large.

ELSE SLIP is Small.

Then the context SLIP is determined by fuzzy-logic processing of the membership grades for τ_s and ℓ_s using *sup-min* connectives. Once the context information is determined through information preprocessing in this manner, the main knowledge base is executed by the compositional rule of inference to compute the inferences for system adjustment (or tuning). These inferences are obtained in the form of membership functions, and are defuzzified by the centroid method. Note that the sub-rule base (Equation (30)) performs a preprocessing function. Its output is the fuzzy quantity SLIP, whereas the inputs are the crisp quantities τ_s and ℓ_s. As expected from the analysis of the previous section, clearly the fuzziness of the information has increased through preprocessing.

Tuning Actions

Adjustment of the conveyor speed and the vertical position of fish is quite straightforward. The fish hold-down control may involve a simple increase or decrease of the holding pressure. The tuning of the controllers (cutter and conveyor) will involve more than one adjustment as, for example, in tuning the PID parameters (de Silva, 1991b). Then, additional submodules of the knowledge base must be developed to represent the associated tuning rules, as given in Figure 13.

Some Results

The performance of the workcell may be judged by analyzing the operation with respect to the objective parameters such as cutting quality, cutting accuracy, and conveyor speed (determines the throughput rate). These parameters are related to the available sensor signals because it is these signals that provide the necessary context for the tuning knowledge base. In particular, let us consider the blade load profile. Four representative results are shown in Figure 16. The load profile in Figure 16(a) was obtained at 80% of the operating speed and under asymmetric loading, in which the top blade was subjected to a significantly higher load than the bottom blade. The reasons range from asymmetric feeding into the cutter to nonuniformity of the fish cross-section. At 120% of the operating speed, the asymmetry was reduced, but the fish was not held correctly. This error resulted in an interference of the fish head with the top blade, as indicated by the second peak at the end of cutting [Figure 16(b)]. At 80% of the operating speed, a satisfactory cut was obtained as clear from the result shown in Figure 16(c), but due to improper holding, some interference with the top blade was present at the end of the cut. Figure 16(d) corresponds to a satisfactory cut at 80% of the operating speed, and this represents a reasonably tuned system.

■ SUMMARY

This chapter discussed a supplemental knowledge-based *hierarchical structure* for a control system. First, the concepts of *fuzzy resolution, degree of fuzziness,* and *control bandwidth* were introduced in the context of the hierarchical control structure. Some analytical representations and illustrations were given. Generally, the fuzziness of information tends to increase, and the fuzzy resolution and the control bandwidth tend to decrease with the level of hierarchy within a control system. A hierarchical model was examined for a fuzzy

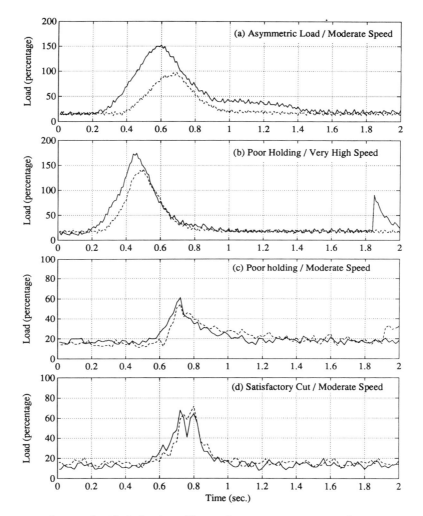

Figure 16 Blade load profiles under various cutting conditions.

logic control system. It contained knowledge systems in upper layers and the knowledge was expressed and processed using fuzzy logic. The model used *prefilters* or *preprocessors* to convert information from lower levels of the hierarchy into context data for a higher level knowledge base. Transitional and combinational operators were involved here. These operations were discussed, particularly with respect to their effect on the fuzziness of information. A fish processing workcell was presented as an example system to which the hierarchical model could be applied. Some results of a practical implementation of the system were given.

■ PROBLEMS

1. One may argue that it is not quite correct to talk about an "intelligent sensor". Instead, one may speak of an "accurate" sensor which faithfully reproduces the measurand (the signal that is being measured). It is what one does with the sensed data that could make the device intelligent. In this sense, it is more appropriate to treat a transducer, or a preprocessor, or a prefilter, or an interpreter as an intelligent device. Discuss this issue and redefine an intelligent sensor so as to remove this inconsistency.

2. Consider a knowledge-based hierarchical control system, that has a "knowledge-based supervisor" which monitors and controls the hierarchies. The layers of the hierarchical structure are organized according to the *level of abstraction* of the processed information and the required speed of response.
   (a) Identify advantages of such a control structure.
   Consider a feedback path from Level i to Level j such that $j > i + 1$. Denote this type of feedback as "Feedback Structure A". Now, suppose that this feedback path is broken into two subpaths as follows:

   Lower path : From i to m ; $i < m < j$
   Upper path : From m to j
   Denote this type of feedback as "Feedback Structure B".
   (b) Discuss advantages and disadvantages of the two feedback structures.

3. A specific sensor may provide several different abstractions of information corresponding to different layers of a hierarchical control system. Give an example of a sensor of this type and briefly describe its use.

4. Consider a hierarchical architecture of a flexible manufacturing system and only the real-time tasks. A hierarchical control structure can be developed where the supervisory (monitoring) level occupies the very top. The lowest level may constitute a set of services that are independent. What are typical functions of the supervisory level in this control structure?

5. Consider a control rule of the form:

   If A and B and .... then C

   Suppose that the membership function $\mu_C(c)$ of the control variable C is triangular with support interval $[-d, d]$ and a unity peak at the mid-point of the support interval (i.e., zero). For a given response condition (context), suppose that the combined membership grade of the context (say, using the *min* operation on the individual grades) is *m*. If the *dot* operation is used, instead of the *min* operation, for obtaining the membership of the corresponding control action C', we have:

$$\mu_{C'}(c) = m\mu_C(c)$$

Obtain an expression for the fuzziness of C', using the 1/2-cut definition. Show that this fuzziness is highest at $m = 1/\sqrt{2}$.

6. Consider an automated process of grading (i.e., quality assessment) of a food product. (As a parallel problem, consider how the "beauty" of a person is determined by a human on the basis of various features). Sensory means are used to obtain information that will establish such features as shape, size, weight, color, texture, and firmness. A rule-based system is used, with these features as inputs, to infer a "grade" (or level of quality) for each object. Comment on the fuzziness and information resolution of various levels of information in this process; particularly that of:
   (a) Raw sensory information.
   (b) Preprocessed information on object features.
   (c) Knowledge-based decision concerning product grade or quality.

7. Learning may be interrupted as repeated preprocessing of information, with some allowance for forgetting (or depreciation). In each cycle of preprocessing, the context may improve further and knowledge will be gained. One simplified model for learning (de Silva, 1993b) is schematically shown in Figure 17. Here, *knowledge* is interrupted as *structured information*. Then *expertise* is interpreted as *specialized knowledge*, and may represent in-depth knowledge that is needed to handle specialized situations. In this model, *intelligence* is placed at the very top of the hierarchy, and may be treated as a high-level controller. It assists in gaining knowledge and expertise. Intelligence itself is fortified through the acts of preprocessing (learning) and could make subsequent stages of learning easier. No attempt is made to give a precise definition for intelligence, in this context, except to state that it represents the

266 INTELLIGENT CONTROL: FUZZY LOGIC APPLICATIONS

capacity to acquire and apply knowledge in an "intelligent" manner. Critically analyze this model.

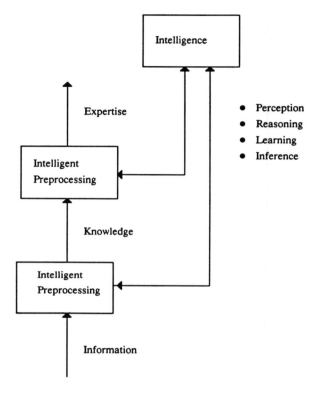

Figure 17 A simplified model of learning

8. Recent research shows that learning words and learning grammar are two quite different activities handled by different parts of the human brain. Can we consider this as a hierarchical problem where, learning words is done at a lower level of the hierarchy, needing less intelligence than learning grammar?

9. Figure 18 shows a schematic representation of a knowledge-based system for quality assessment (grading) of processed herring roe (skeins). It uses preprocessed sensory information on shape, size, color and texture, as obtained from CCD cameras, ultrasound probes, etc. Expert graders are available who will provide grading criteria, and will serve as domain experts. Outline the process of developing a knowledge-based system for this application. Give a simplified rule base that can be implemented using fuzzy logic in this problem.

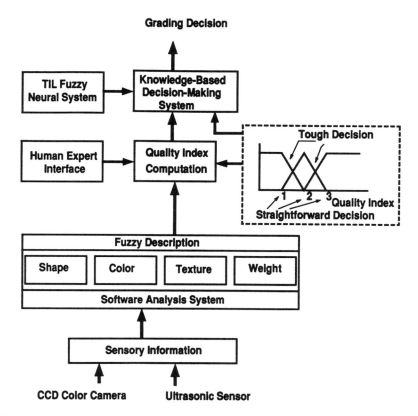

Figure 18 Schematic representation of a knowledge-based system for grading herring roe skeins.

▬ REFERENCES

Albus, J.S., McLean, C., Barbera, A., and Fitzgerald, M. (1982). An Architecture for Real-Time Sensory-Interactive Control of Robots in a Manufacturing Environment, *Proc. 4th IFAC/IFIP Symp. Information Control Problems in Manufacturing Technology*, Gaithersburg, MD.

de Silva, C.W. (1989). *Control Sensors and Actuators*, Prentice-Hall, Englewood Cliffs, NJ.

de Silva, C.W. (1990). Research Laboratory for Fish Processing Automation, *Robotics and Manufacturing*, Vol. 3, pp. 935–940, ASME Press, New York.

de Silva, C.W. (1991a). Fuzzy Information and Degree of Resolution within the Context of a Control Hierarchy, *Proc. IECON'91*, Kobe, Japan, pp. 1590–1595.

de Silva, C.W. (1991b). An Analytical Framework for Knowledge-Based Tuning of Servo Controllers, *Int. J. Eng. Appl. Artif. Intelligence*, Vol. 4, No. 3, pp. 177–189.

de Silva, C.W. (1993a). Hierarchical Preprocessing of Information in Fuzzy Logic Control Applications, *Proc. 2nd IEEE Conf. Control Applications*, Vancouver, Canada, Vol. 1, pp. 457–461.

de Silva, C.W. (1993b). Knowledge-Based Dynamic Structuring of Process Control Systems, *Proc. 5th Int. Fuzzy Systems Assoc. World Congr.*, Seoul, Korea, Vol. II, pp. 1137–1140.

de Silva, C.W. (1995). Considerations of Hierarchical Fuzzy Control, in *Theoretical Aspects of Fuzzy Control*, Nguyen, H., Sugeno, M., Tong, R., and Yager, R.R. (Eds.), John Wiley & Sons, New York.

de Silva, C.W. and MacFarlane, A.G.J. (1988). Knowledge-Based Control Structure for Robotic Manipulators, *Proc. IFAC Workshop on Artificial Intelligence in Real-Time Control*, Swansea, U.K., pp. 143–148.

de Silva, C.W. and MacFarlane, A.G.J. (1989). *Knowledge-Based Control with Application to Robots*, Springer-Verlag, Berlin.

de Silva, C.W. and Wickramarachchi, N. (1992). Knowledge Base Development for an Intelligent Control System of a Process Workcell, *Intelligent Control Systems*, DSC-Vol. 45, Winter Annual Meeting, ASME, New York, pp. 113–119.

Dubois, D. and Prade, H. (1980). *Fuzzy Sets and Systems*, Academic Press, Orlando, FL.

Isik, C. and Meystel, A. (1986). Decision Making at a Level of a Hierarchical Control for Unmanned Robot, *Proc. IEEE Int. Conf. Robotics and Automation*, IEEE Computer Society, Los Angeles, Vol. 3, pp. 1772–1778.

Riahi, N. and de Silva, C.W. (1990). Fast Image Processing for the Gill Position Measurement in Fish, *Proc. IECON'90*, Pacific Grove, CA, Vol. 1, 90CH2841-5, pp. 476–481.

Saridis, G.N. (1988). Knowledge Implementation: Structures of Intelligent Control Systems, *J. Robot. Syst.*, Vol. 5, No. 4, pp. 255–268.

Zadeh, L.A. (1975). The Concept of a Linguistic Variable and its Application to Approximate Reasoning, Parts 1-3, *Inf. Sci.*, pp. 43–80, 199–249, 301–357.

Zadeh, L.A. (1979). A Theory of Approximate Reasoning, in *Machine Intelligence*, Hayes, J. et al. (Eds.), Vol. 9, pp. 149-194.

9 INTELLIGENT RESTRUCTURING OF PRODUCTION SYSTEMS

INTRODUCTION

Automation is vital to industrial processes in achieving increased productivity and high efficiency, with associated economic benefits. In a flexible production system (FPS) the process components are typically organized as distinct workcells in which each workcell is dedicated to a specific class of production activity. Even though the individual workcells are flexible in the sense of easy reconfiguration to accommodate changes in product types and batch sizes, the overall architecture and the constituent components of a conventional workcell are kept fixed (Weatherhall, 1988; Kovacs and Mezgar, 1991). Changes in production demand, variations in raw material supply, and malfunction or failure of workcell components can increase or decrease the workload on some components in the workcell. Overloaded workcell components will result in degraded system performance. Also, components that operate significantly below their full capacity will cause reduced productivity and degraded system efficiency. It is clear that for optimal performance, the workcell components should operate uniformly close to their full capacity.

The concept of dynamic sharing of workcell components has been proposed as an approach to optimizing the operation of a FPS (de Silva, 1993a,b). The idea is to continuously monitor the workcell components and systematically shed overloading through sharing those components that operate below capacity with those that are overloaded, assuming that the components shared in this manner are sharable or interchangeable. Such sharing cannot be done arbitrarily. A thorough knowledge of the FPS is necessary in matching an undercapacity component with an overloaded one, for the purpose of load sharing. Past experience and knowledge alone would be inadequate. Consequently, an upper-level knowledge-based system would be desirable to make the component sharing decisions, based on low-level sensory information that is obtained from the workcell components. Because this knowledge is "qualitative", at least in part, and may be human based and incomplete, fuzzy logic would be appropriate for both representation and processing of the associated knowledge (Zadeh, 1979; de Silva, 1991).

A theoretical framework is presented in this chapter for the dynamic sharing of workcell components (de Silva and Gu, 1994). A knowledge-based approach that employs fuzzy logic is given for decision making related to component sharing. The method depends on a systematic procedure for matching an overloaded component with an undercapacity component of the same sharable type. Next, a blackboard architecture is given for implementation of the approach in Prolog. A case study is described and computer simulations are presented to illustrate the application of the approach in a realistic industrial process.

■ THEORETICAL FRAMEWORK

The conceptual development of the approach presented in this chapter is based on the theoretical framework that is outlined in this section (de Silva and Gu, 1994). Consider the control and communication architecture of a FPS (de Silva, 1993b), as shown in Figure 1. The FPS consists of several workcells (WC), with two-way communication between each cell-host controller and the FPS controller via a high-speed local-area network (LAN). A workcell consists of several components (c), which may include processing tools (e.g., manipulators, grippers, positioners, integrators, and disintegrators) and monitoring tools (de Silva, 1989) (e.g., sensors, vision stations), with associated component controllers. Note here that the term "integrator" is used generically to denote devices for such processes as welding, gluing, sewing, riveting, and fastening. Similarly, the term "disintegrator" is used to denote the processing tools such as cutters, lathes, milling machines, and saws. The workcell components interact with the process environment, including workpieces, in carrying out the production task. Such interactions must be monitored through appropriate sensors and the acquired information must be provided to the component controllers at high resolution, and may be properly preprocessed and provided to the workcell controllers and the FPS controller at a lower resolution (de Silva, 1991).

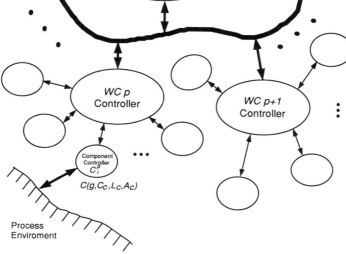

Figure 1 Control and communication architecture of a flexible production system.

Hierarchical Structure

A hierarchical control structure (de Silva and MacFarlane, 1989) is assumed for the overall system, as shown in Figure 2. Three distinct layers are shown. At the lowest level, the workcell components are grouped such that each group (g) contains a set of components that may be shared or interchanged to some degree among the workcells within the overall FPS. The sensory signals from each component are used by the component controller for direct, local control. This high-resolution information may be preprocessed, possibly using "intelligent" preprocessors, to provide compatible low-resolution information for the upper layers of the control hierarchy (de Silva, 1991). A workcell controller performs higher level control tasks than those of the component-level controllers. For example, component coordination, generation of the reference signals to drive the components, downloading of component-task programs, and monitoring of the workcell components are all done by the workcell controllers. Some intelligent control activity might be needed at this intermediate level as well. This would require a *knowledge system* within the layer. The context information for the knowledge system could come from preprocessed sensory information and also from upper layers and through operator interfaces.

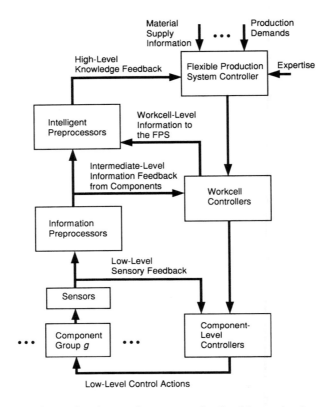

Figure 2 A hierarchical control structure of a flexible production system.

The top layer of the three distinct layers shown in Figure 2 corresponds to the FPS controller. Task planning, procedural decomposition, subtask allocation, and overall system monitoring all may be carried out at this level. This typically is the "most intelligent" controller in the hierarchy. Its knowledge system may have to deal with qualitative, incomplete, or vague information and may encounter unplanned or unfamiliar situations. It may depend on human expertise, heuristics, past experience and learning. The knowledge system may be driven in part by the low-level sensory information, subjected to intelligent preprocessing, as schematically shown in Figure 2. The intelligent preprocessing would provide low-resolution, high-level information, and could be interpreted as a form of perception or "intelligent sensing" of the system status. The FPS controller relies on many forms of external data, expertise, and knowledge.

Restructuring Problem

Consider the multiple-workcell FPS that is schematically shown in Figure 1. The pth workcell is denoted by WC_p and its ith component is denoted by c_i^p. For simplicity, the subscript and superscript of this component notation are dropped in the sequel, except when they are explicitly needed. Suppose that the workload demand for a particular workcell, as determined by the FPS controller, is D_w. The corresponding workloads L_c of the individual components of this workcell may be determined using a standard task planning procedure. This procedure may be expressed as the nonlinear relationship

$$L_c = L_c(D_w) \qquad (1)$$

Suppose that the processing capacity of a workcell component c is C_c. Also at a given instant, suppose that the production level of a workcell is P_w and the activity level of component c of this workcell is A_c. Then, the following relations hold:

$$\text{IF } L_c \leq C_c \text{ for } \forall c \varepsilon W$$
$$\text{THEN } A_c = L_c \qquad (2)$$

$$\text{IF } \exists \text{ some } c \varepsilon W \text{ such that } L_c > C_c$$
$$\text{THEN } A_c \leq C_c \qquad (3)$$
$$\text{AND } P_w < D_w$$

Here, W denotes the set the components in a particular workcell.

Next, consider a group g of "similar" components which are sharable or interchangeable within the group. Then, the characteristics of a component c may be represented by the ordered set [g, C_c, L_c, A_c], as shown in Figure 1. Suppose that at a given instant of workcell operation, the workcell components are monitored and some are determined to operate below capacity (undercapacity), while some others are overloaded. Then, the following conditions hold:

$$\text{IF } c \varepsilon g \text{ AND } L_c < C_c \text{ THEN } c \varepsilon g_u$$
$$\text{IF } c \varepsilon g \text{ AND } L_c > C_c \text{ THEN } c \varepsilon g_o \qquad (4)$$

with

$$g_u \cap g = g_u \qquad g_o \cap g = g_o$$

where

g_u = subgroup which contains all components that operate below capacity (undercapacity) within the component group g

g_o = subgroup which contains all components that are overloaded within g

Note that in general, $g_u \cup g_o \subseteq g$, because for some components within g it is possible that $L_c = C_c = A_c$. Also, even when $L_c = C_c$ for a particular component, it is possible that $A_c < C_c$ for that component as a result of gross reduction of the production level P_w of the workcell that contains the component, due to an overload in some other component within the same workcell. Also, in the foregoing formulation, the variables are assumed to be appropriately normalized so that direct comparison is possible.

The optimal restructuring problem may be expressed as

$$\text{minimize } J = J(\alpha, \mathbf{C}, \mathbf{A}, \mathbf{L})$$

$$\text{subject to } P_w(\mathbf{A}) = D_w(\mathbf{L}) \tag{5}$$

where α is a vector of weighting parameters, and the cost function J is defined for the entire FPS. Also, $\mathbf{C}$, $\mathbf{A}$, and $\mathbf{L}$ are vectors of capacities, activities, and loads of components within the FPS. For example, a quadratic cost function could be used; thus

$$J = \sum_c \alpha_c (C_c - A_c)^2 \tag{6}$$

with nonsymmetric weighting parameters such that

$$\alpha_c = \alpha_{co} \quad \text{if} \quad L_c > C_c$$

$$\alpha_c = \alpha_{cu} \quad \text{if} \quad L_c \leq C_c$$

$$\alpha_c = 0 \quad \text{if} \quad A_c = 0 \tag{7}$$

It should also be noted that no activity (zero activity or $A_c = 0$) is considered a desirable condition. In other words, rather than having, e.g., two undercapacity components of the same type, it is desirable to transfer the load of one component into the other and completely release the first component. This is advantageous for reasons such as reduction in operating costs and wear and tear. Furthermore, the component that is released of activity in this manner will be available for absorbing overloads from other similar components in the FPS. This aspect is addressed later in the chapter. It is clear that zero-activity components do not contribute to the cost function J, as expressed by Equation (6). Accordingly, the optimization strategy inherently favors the existence of zero-activity components.

Physically, the restructuring problem may be interpreted as scanning of every component group g separately, and instituting a load sharing strategy so that if successful, no overloaded components would exist ($g_o = \phi$ for $\forall g$). Furthermore, no undercapacity component would exist that could be released by transferring its load to another without causing an overload. For such strategy to be successful, the components may have to be

1. Shared within a workcell
2. Shared between several workcells
3. Moved from one workcell to another

In the first case the workcell structure must be flexible so that it can be modified. In the latter two cases the overall system structure will have to be modified. This is known as dynamic structuring (de Silva, 1993a,b).

Knowledge-Based Restructuring

Restructuring of components in a multiple-workcell FPS should not be carried out using purely analytical optimization criteria alone. Certain factors or conditions may favor some structures, while objecting to some others. The factors that determine the desirability of a particular structure of component sharing may be based on nonanalytic criteria such as knowledge and history (e.g., repair state and reliability) of a particular component, experience in the operation of the FPS, expert opinion, and heuristics. Under such conditions, a knowledge-based approach to making the restructuring decisions would be desirable.

Consider a group g of sharable or interchangeable components which are associated with a set of workcells **w**. Each workcell in this set has an associated set of components $\mathbf{c} \subset g$. Then the knowledge-based restructuring process may be expressed as

$$g(\mathbf{w}^*(\mathbf{c}^*)) = R(g) \otimes g(\mathbf{w}(\mathbf{c})) \quad \text{for all } g \text{ within the FPS} \quad (8)$$

Here, $R(g)$ may be interpreted as a knowledge system which evaluates the present association of components **c** with workcells **w** within a group g, and then modifies the associations appropriately so as to reduce component overloading. The resulting new association of components $\mathbf{c}^*$ within g, now with a set of modified workcells $\mathbf{w}^*$, is denoted by $g(\mathbf{w}^*(\mathbf{c}^*))$. Note that the flexible production system is said to be restructured when the constitution of the workcells changes due to such modified association of components. The decision-making process given by Equation (8) must be repeated for all groups of components within the FPS.

Again, consider a group of sharable or interchangeable components g, with m overloaded components represented by the subgroup g_o and r undercapacity components represented by the subgroup g_u. Note that the set cardinalities satisfy the relationship

Context Levels		Decision Levels
Overloaded Component OC	Undercapacity Component UC	Matching Priority PR
	No activity na	
Highly overloaded ho	Highly undercapacity hu	Very high vh
Moderately overloaded mo	Moderately undercapacity mu	High hi
		Moderate md
Lightly overloaded lo	Lightly undercapacity lu	Low lw

Rulebase:

```
       If  OC is ho and UC is na then PR is vh
Else   if  OC is ho and UC is hu then PR is vh
Else   if  OC is ho and UC is mu then PR is hi
Else   if  OC is ho and UC is lu then PR is md
Else   if  OC is mo and UC is na then PR is hi
Else   if  OC is mo and UC is hu then PR is hi
Else   if  OC is mo and UC is mu then PR is md
Else   if  OC is mo and UC is lu then PR is md
Else   if  OC is lo and UC is na then PR is lw
Else   if  OC is lo and UC is hu then PR is lw
Else   if  OC is lo and UC is mu then PR is md
Else   if  OC is lo and UC is lu then PR is md
End    if.
```

Figure 3 The rule base of component matching for load sharing.

$$\text{card}(g) \geq \text{card}(g_o) + \text{card}(g_u) \quad (9)$$

In knowledge-based restructuring each *overloaded component* (OC) must be matched with an appropriate *undercapacity component* (UC). First, a *priority* (PR) is determined and assigned for a given match. This may be accomplished using the rule base given in Figure 3. The states that a component may take are given by {no activity, undercapacity, balanced, overloaded}. Even though the condition of "no activity" (na) may be treated as a special case of "high undercapacity" (hu), a separate state has been defined in Figure 3 for this condition for reasons that should be clear later in the context of component releasing. In particular, the state of na is far more desirable than that of hu. A balanced state is said to exist in a component when its activity is equal to its capacity (i.e., $A_c = C_c$). This is the optimal state of component operation. The membership functions of the fuzzy resolution levels for the two context variables OC and UC and for the decision variable PR should be known in using fuzzy logic to represent and process (de Silva, 1991) the knowledge base given by Figure 3. The compositional rule of inference (Zadeh, 1979; Dubois and Prade, 1980) is employed for decision making with this fuzzy logic rule base (see Chapters 3 and 4).

In prioritizing the load sharing decisions the factors considered in the rule base of Figure 3 are the level of overload $L_{co} - A_{co}$ of the overloaded components and the level of undercapacity $C_{cu} - A_{cu}$ of the undercapacity components. This information is "crisp", as obtained from sensory data of the workcell components, and must be fuzzified using the corresponding membership functions in the process of applying the compositional rule of inference. Furthermore, the priority decisions obtained through this process are themselves fuzzy and given by the decision membership functions. They must be defuzzified, for example, using the *centroid method* (see Chapters 3 and 4), in order to obtain crisp priority values.

Apart from the overload level $L_{co} - A_{co}$ and the undercapacity level $C_{cu} - A_{cu}$ there are several other factors that should be taken into consideration in making a final decision for load sharing. These factors include ease of sharing as determined, for example, by the proximity of the components, the degree of reliability and shareability of the undercapacity component that is being considered for load sharing, and other subjective information that may be available through past experience (e.g., cost of operation, the state of repair, age, design life) and expert opinion. For example, all other factors being equal, one should favor the component with least operating cost per unit task in making the decision of component sharing. Such information may be expressed as a matrix of *feasibility indices*:

$$\mathbf{F} = [f_{ij}]_{m \times r} \tag{10}$$

where m is the number of overloaded components and r is the number of undercapacity or no-activity components in group g. This matrix may be generalized, with $m = r$ = total number of components in g. In this case $\mathbf{F}$ is a square, but not necessarily symmetric, matrix with zero diagonal elements.

Each defuzzified priority value for a matching pair of components being considered for load sharing must be multiplied by the corresponding feasibility index, as given by Equation (10), in order to arrive at a load sharing decision. The matrix size and the parameter values of $\mathbf{F}$ must be suitably updated following each decision of load sharing.

Using the conventional *max-min* or *max-dot* composition in fuzzy logic decision making, the *priority value* $\widehat{PV}$ for sharing an overloaded component *OC* with an undercapacity component *UC* may be computed using the relations given below. First, the membership function of the priority value *PV*, which is represented as a fuzzy descriptor, is determined by

$$\mu_{PV}(y) = \max_{\substack{i,j \\ k(i,j)}} \{\min[\mu^i_{oc}(L_{oc} - A_{oc}), \mu^j_{uc}(C_{uc} - A_{uc})] \cdot \mu^k_{PR}(y)\} f(OC, UC)$$

(11)

where

$\mu^b_A(\cdot): \Re \to [0,1]$ = membership function of the *b*th fuzzy resolution of the fuzzy descriptor *A*, which is a function mapped from the real line $\Re$ onto the closed real interval [0,1]

f(c,d) = feasibility index of sharing an undercapacity component *d* with an overload component *c*

Next, a crisp value $\widehat{PV}$ for the sharing priority is obtained using the centroid method, as

$$\widehat{PV} = \frac{\int_{\overline{PR}} y\mu_{PV}(y)dy}{\int_{\overline{PR}} \mu_{PV}(y)dy}$$

(12)

where $\overline{PR}$ is the support set (interval) of the fuzzy set *PR* in Equation (11).

Component Releasing

As noted above, an undercapacity component (or a no-activity component) may be shared with an overloaded component within the same workcell (case 1) or within another workcell (case 2). In another scenario (case 3), first, two undercapacity components within the same workcell are compared and the load from one component is transferred to the other, with the possibility of completely releasing the load of the first component. Then, the resulting zero-activity (no-activity) component may be shared with another component as in case 2 (or even case 1, although less likely). Even though this zero-activity component is in fact a special case of a highly undercapacity (*hu*) component for the purpose of load sharing with an overloaded component, a distinction should be made in this regard during the process of component releasing. Otherwise, infinite looping could result here, with the load being transferred back and forth between an undercapacity component and a zero-activity component. Furthermore, an undercapacity component need not be completely released in the intermediate load-transfer process. For example, a lightly undercapacity (lu) component might be converted into either a moderately undercapacity (mu) component or a highly undercapacity (hu) component in this cycle, and would be ready for sharing with an overloaded component (regardless of whether ho, mo, or lo) in the next cycle of load transfer.

The rule-based approach to load transfer between two undercapacity components is somewhat similar to that between an overloaded component and an undercapacity component, but the rule base that is used would be quite different from what is given in Figure 3. As we have noted, no zero-capacity components would be involved now. Consider an undercapacity component (UCR) that is expected to be released of its load, and a second undercapacity component (UCL) that is expected to receive the load of the first component. An appropriate rule base for determining the priority for this load transfer is given in Figure 4. Here, too, a feasibility matrix of load transfer is used, and the application of the compositional rule of inference would be given by a counterpart of Equation (11), as

$$\mu_{PV}(y) = \max_{\substack{i,j \\ k(i,j)}} \{\min[\mu^i_{UCR}(C_{UCR} - A_{UCR}), \mu^j_{UCL}(C_{UCL} - A_{UCL})] \cdot \mu^k_{PR}(y)\} f(UCR, UCL)$$

(13)

The subscripts L and R have been added consistently to denote the releasing component and the loading component, respectively. The priority value $\widehat{PV}$ is computed as before using centroidal defuzzification, as given by Equation (12).

A general system for knowledge-based component sharing is schematically shown in Figure 5. Note that the system allows for updating the feasibility matrix after each decision-making stage. It also allows for selecting and updating the rules of component matching for the process of load sharing. The main steps associated with the knowledge-based load sharing (and load transfer) process are given below:

Step 0: Initialize the rule base, membership functions, feasibility index matrix, component capacities, and groups of sharable component.
Step 1: For the present level of production demand, determine the component loads.
Step 2: From sensory information, determine the component activity levels. Also, update the feasibility index matrices, and if necessary, the component capacities.
Step 3: Compute component states, including overloads and undercapacities.
Step 4: In the next two steps carry out a component-releasing cycle by comparing pairs of undercapacity components within the group, but excluding the no-activity components.

Context Levels		Decision Levels
Releasing Undercapacity Component UCR	Loading Undercapacity Component UCL	Load Transfer Priority PR
Highly undercapacity hu	Highly undercapacity hu	Very high vh
Moderately undercapacity mu	Moderately undercapacity mu	High hi
		Moderate md
Lightly undercapacity lu	Lightly undercapacity lu	Low lw

Rulebase:

```
      If    UCR  is  hu  and  UCL  is  hu  then  PR  is  hi
Else  if    UCR  is  hu  and  UCL  is  mu  then  PR  is  vh
Else  if    UCR  is  hu  and  UCL  is  lu  then  PR  is  hi
Else  if    UCR  is  mu  and  UCL  is  hu  then  PR  is  md
Else  if    UCR  is  mu  and  UCL  is  mu  then  PR  is  md
Else  if    UCR  is  mu  and  UCL  is  lu  then  PR  is  md
Else  if    UCR  is  lu  and  UCL  is  hu  then  PR  is  lw
Else  if    UCL  is  lu  and  UCL  is  mu  then  PR  is  lw
Else  if    UCR  is  lu  and  UCL  is  lu  then  PR  is  lw
End   if.
```

Figure 4 The rule base for load transfer from one undercapacity component to another.

Step 5: Apply the compositional rule of inference with the context information for appropriate pairs of components, and defuzzify the load transfer priorities in each group. Apply the feasibility indices.

Step 6: Select the highest priority pair of components. Institute the load transfer action.

Step 7: If a component releasing cycle was just completed, go to step 5.

Step 8: If either no overloaded components exist, or further improvement in load distribution is not possible, then idle. Otherwise proceed to the next group and go to step 1.

In the subphase of component releasing a no-activity component ($A_c = 0$) should not be treated in the same category as an undercapacity component. Otherwise, infinite looping of load transfer could result during this stage of component releasing. Specifically, a load transfer from an undercapacity component into another undercapacity component could release the first component, while keeping the second component below its capacity. In the next cycle the load might be transferred from the second component back into the first component, and so on. Also, complete elimination of

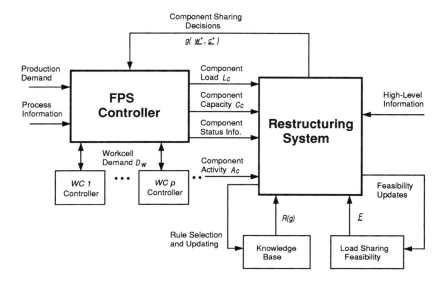

Figure 5 A knowledge-based scheme for component sharing in a flexible production system.

overloading might not be realizable with the existing system resources and for the present process demand. Thus, termination of the restructuring process would occur even while some of the components remain overloaded.

■ IMPLEMENTATION USING A BLACKBOARD ARCHITECTURE

The knowledge-based framework for component sharing, which was developed in the previous section, may be implemented using a blackboard model (Englemore and Morgan, 1988; Jaganathan et al., 1989). A particular layer of a knowledge-based system may take the blackboard architecture. This architecture consists of a global (shared) data region called *blackboard* (BB), several *knowledge sources* (KS) or intelligent modules that interact with this data region, and a *control unit*. The knowledge sources are not arranged in a hierarchical manner and will cooperate as equal partners (specialists) in making a knowledge-based decision. The knowledge sources interact with the shared data region under the supervision of the control unit. When the data in the blackboard change, which corresponds to a change in context, the knowledge sources would be triggered in an opportunistic manner and an appropriate decision would be made. That decision could then result in further changes to the blackboard data and subsequent triggering of other knowledge sources. Note that the data may be changed by external means (for example, through a user interface) as well as due to knowledge–source actions.

The decision-making process of the restructuring problem that was presented in the previous sections is essentially opportunistic, in which an appropriate partition (or a knowledge source) of the overall knowledge base would act at a given time in carrying out the most appropriate action (e.g., information updating, load planning, restructuring). It follows that a blackboard model would be suitable for the implementation of the associated knowledge-based system. Furthermore, the following characteristics of the problem and its implementation should be noted:

- The knowledge sources may take different forms. They may include rule-based decision making, planning procedures, information processing, or even look-up table updating. In a blackboard architecture the knowledge sources can be implemented as independent modules, with a common data region (blackboard), thereby avoiding complicated data passing.
- The practical control mechanism of the problem described in the second section can be quite complex. For example, after the system makes a sharing decision, the control may continue with another decision if an overload remains, may determine component loads if the task demands change, or may re-establish the system activity level, if some components become disabled. The blackboard architecture provides a mechanism of opportunistic control, which can carry out such steps in the present application.
- The FPS information should be accessible by every knowledge source, in which the integrated data of the overall operation of the FPS are needed for the reasoning process. Such data may be posted on a blackboard which would be visible to all knowledge sources. Also, all the information changes are posted on the blackboard.

The implementation of the present problem is now addressed in the context of both the architecture of the overall system and the planning strategy, with particular emphasis on the decision making associated with planning (de Silva and Gu, 1994).

System Architecture

The blackboard system architecture used in the implementation of the restructuring problem of a FPS is shown in Figure 6. The shared data region (blackboard) is partitioned into five subregions (sub-blackboards) denoted as "Task Demand", "Component Information", "Workcell Information", "Sharing Feasibility", and "Restructuring Records". Also, there are six knowledge sources dealing with information updating and restructuring. A control unit supervises the blackboard and triggers a proper knowledge source to solve the problem associated with the particular context (data change). This structure is explained below.

INTELLIGENT RESTRUCTURING OF PRODUCTION SYSTEMS 283

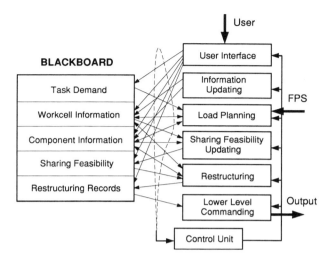

Figure 6 Blackboard structure of the restructuting system.

Domain Blackboard

Domain knowledge of a FPS structure, including operating conditions, is represented as objects in the shared data area (blackboard). An object may be generated by its frame (or template or schema). For example, the frames of "component" and "workcell" take the following structure:

    Component: **Workcell:**
     group;   *type;*
     position;   *demand;*
     work_for;   *activity_level;*
     capacity;   *configuration;*
     work_status;   *layout.*

Note that each of these two objects has five properties, as listed above. The value of a property may be expressed in the form of Prolog predicates:

$$\text{component } (N, P, V)$$

$$\text{workcell } (N, P, V)$$

The first predicate denotes that for the component N, the value of the property P is V. Similarly, the second predicate gives the value V of property P of the workcell N. These two may be combined into a single predicate, for example, *holds(N,P,V)*, but for clarity, two separate predicates are used here. The values of these objects are stored in the corresponding sub-blackboard areas. In this example the

"Component" objects are stored in the "Component Information" region of the blackboard, and similarly, the "Workcell" objects are stored in the "Workcell Information" region.

Similar explanations apply to three other regions of the blackboard shown in Figure 6. For example, the values of "Task" objects, which are stored in the "Task Demand" region of the blackboard, may be specified as the type of task, an associated property, and a corresponding value. Three examples that are related to an automated fish processing plant, as described in the next section, are given below:

> task (cutting, demand, 1360 ton/day).
>
> task (packaging, demand, 1000 ton/day).
>
> task (grading, demand, 1000 ton/day)

Note that the "task demand" given here is not necessarily identical to the "workcell demand". For instance, this is the case in which the components of more than one workcell contribute to accomplishing a particular task. Also, note that the information on "Sharing Feasibility" is specified for pairs of components of the same type. The values of these various objects on the blackboard are changed by the knowledge sources that interact with them. This aspect is addressed next.

Knowledge Sources

Knowledge sources are designed independently of each other, and consist of the knowledge base modules for various activities of the system. In the architecture shown in Figure 6 there are six KSs that deal with user interface, information updating, load planning, updating of sharing feasibility, restructuring, and output. Main characteristics of these knowledge sources are outlined below:

- **KS of user interface.** This handles the input of production demands. It also provides a function that enables a user to change the information in the blackboard. It has the highest priority without any precondition.
- **KS of information updating.** This KS updates the operating information of the FPS, including the available capacity resources, workcell activity levels, and the system layout. It obtains data from the component layer through information preprocessors and from the workcell layer through intelligent preprocessors. This KS has the second priority.
- **KS of load planning.** This KS has a knowledge base for capacity planning methods and associated transformations that determine component loads for a specified demand [see Equation (1)]. It assigns the workload to the workcell components according to the

INTELLIGENT RESTRUCTURING OF PRODUCTION SYSTEMS   285

processing demand and the task planning method, and then sets the component status according to component capacity, load, and the activity level. This KS can have several options for assigning loads to components of different capacity levels, which will affect the outcome of the restructuring process. Then, it can automatically select another option if a restructuring process fails. This KS has the third priority.

- **KS of updating the sharing feasibility.** This KS obtains information about component shareability (including considerations of component reliability and operating cost) and geographic factors (including distance between components), and updates the sharing feasibility. Because this should be done before restructuring, it has a higher priority than that for KS of restructuring.

- **KS of restructuring.** This KS performs a procedure of restructuring planning according to its knowledge base employing fuzzy logic. A planning method used for this purpose is presented later in the chapter.

- **KS of commanding lower level.** This sends restructuring commands to the workcell controllers, which are located at a lower level. Some coordination of the shared components should be done here. In addition, the restructuring commands might have to be approved by system experts. This KS has the lowest priority.

The structure of a knowledge source is of the form

KS :

trigger_status;

priority;

body.

The parameters of a KS may be specified by the Prolog predicate: *ks(N, I, V)*, which states that the item *I* of the KS *N* has the value *V*. For example, the "KS of restructuring" may be expressed as:

ks(restructuring, trigger_status, on).

ks(restructuring, priority, 2).

ks(restructuring, body, planning)

where *planning* is a predicate name for the restructuring planner, which is explained in a subsequent section.

Control Unit

The main purpose of the control unit of a blackboard system is to monitor the blackboard and trigger the appropriate knowledge sources in response to a data change in the blackboard. This may be accomplished according to some priority. For this reason, the control unit is sometimes termed a "sequencer". Note that the data change may be effected either through an external input or by a KS action itself. When the system is in the idle state, the knowledge sources "User Interface" and "Information Updating" should be triggered automatically. A simple control mechanism for the blackboard system can be expressed by:

control ←

    collect_triggered_KS(KSs),

    select_highest_PR(KSs, KS),

    execute (KS),

    control.

control ←

    trigger_user_interface,

    trigger_info_update,

    control.

Here, the operation *collect_triggered_KS(KSs)* collects all triggered knowledge sources. It may fail in case no KS is triggered, which will cause the control to turn to the second clause, which will then trigger the knowledge sources of "User Interface" and "Information Updating". Note that the trigger status of a KS will be removed only after the KS has been executed.

As an example, suppose that the blackboard region "Component Information" has been changed. Then the knowledge sources "Sharing Feasibility Updating" and "Restructuring" will be triggered, and the former, which has the higher priority, will be executed first. This execution may change the BB of "Sharing Feasibility", and can in turn trigger the KS of "Restructuring" again. Its execution may cause some other blackboard regions to change. This will trigger some more knowledge sources and consequently change the data in the associated blackboards, and so on. The procedure will end when an execution of the knowledge sources does not result in further changes of the BB data, and the trigger status of all the knowledge sources are off.

Restructuring Planner

As formulated in this chapter, the purpose of the restructuring process is to optimize the FPS performance (see cost function J in Equations (5) and (6)) by operating the workcell components as close as possible to their capacity levels. The restructuring planner assigns component loads according to some strategy, as described before, and consequently restructures the operation of the system. A blackboard-based restructuring planner is described in this section (de Silva and Gu, 1994).

Restructuring Heuristics

Actions of the restructuring planner may be determined by past experience and expertise. Typical heuristics used in this process would be:

1. Moving a component to the workcell for which it works.
2. Terminating a sharing action if either the sharing feasibility drops below some preassigned threshold value or the component load is no longer above its capacity.
3. Releasing an undercapacity component by transferring its workload activity to other components of the same type within the same workcell.
4. Sharing the load of an overloaded component with other components of the same type within the same workcell.
5. Component releasing as in case 3, but through load transfer between different workcells.
6. Component sharing as in case 4, but through load transfer between different workcells.

The inference mechanism for using the above heuristics is shown in Figure 7, in which the objects denoted by 1, 2, 3, etc., are arranged in the order of the priority of executing the associated rules. Note that only when the rules with higher priority have failed to solve the problem can a lower priority rule be fired. When a rule succeeds, which means the blackboards may receive new data by the actions of the rule, the inference engine will review the entire problem in the blackboard all over again. This process will end when none of the rules of the knowledge sources are triggered.

This mechanism is implemented in Prolog as follows:

planning ←

heuristic (A),

execute (A),

planning.

planning ←

calculate_cost_function (J),

report_planning_result (J).

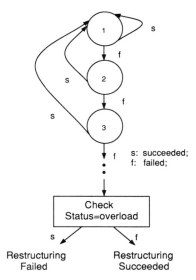

Figure 7 The reasoning sequence.

Here, *heuristic(A)* picks one of the six heuristics itemized above, and selects an action A. Then, the operation *execute(A)* performs the action A. This procedure is repeated until all the heuristics have failed to pick an action. Then the second procedure will calculate the cost function J and report the restructuring results.

Fuzzy Decision Making

A common problem in any knowledge-based system is how to pick an action when more than one action of the same type is available. This is known as *conflict resolution*, and heuristics may be used for this purpose. As discussed previously, a decision-making procedure incorporating fuzzy logic is used for selecting an appropriate action. As an example, consider the following heuristic:

heuristic (A) ←

    group(G),

    group_component(overloaded, G, OL),

    possible_actions(OL, As, within),

    select_highest_PR(As, A, PD),

    PD > 0.3.

This heuristic is used for the action of sharing the load of an overloaded component with an undercapacity component within the same workcell. The operation *group_component(overloaded, G, OL)* groups all the overloaded components of type *G* into a list *OL*, which is then passed on to *possible_actions(OL, As, within)* to find all possible actions *As*. Note that a dynamic database is used in executing *possible_actions*. Another predicate, *priority(A, PR)*, which means action *A* has priority *PR*, is used for storing the available actions with associated priority degrees. All the possible actions are stored and then used for *select_highest_PR(As, A, PD)*, which picks the action *A* with the highest priority *PD* in the action list *As*. Finally, all stored actions are retracted. The flags "overloaded" and "within" can be changed to "under_capacity" and "between" for other possible heuristics; for example, component releasing and load transfer "between" workcells. A threshold should be set for the lowest acceptable assessment value of an action.

In the heuristic given above it was necessary to determine all possible actions ("possible_actions") for load transfer. For simplicity, consider just one overloaded component *OC* instead of an entire list *OL*. The new predicate is named *possible_acts*. Consider the following Prolog implementation:

possible_acts (OC, _, within) ←

    component (OC, position, PO),

    component (UC, position, PO),

    associative (OC $\cap$ UC $\Rightarrow$ A, PR),

    assert (priority (CA, PR)),

    fail.

possible_acts(_, AL, _) ←

    findall (A, priority (A, _), AL).

Here the condition "fail" (the built-in predicate *fail* is used) forces the *associative (OC $\cap$ UC $\Rightarrow$ A, PR)* to find all actions of load sharing between an overloaded component *OC* and possible undercapacity components *UC* of the same type as *OC*, and store them with their corresponding priority degree values *PR*. The second clause terminates the search after a success and returns all possible actions which were found by the first clause.

The context information $OC \cap UC$ of the *associative* $(OC \cap UC \Rightarrow A, PR)$ forms an Euclidean product space of the dimensions of OC and UC. This will be used to fire rules for action A. Note that each rule for a particular layer (see Figures 3 and 4) will contribute to the priority degree of the action. The resulting fuzzy priority degree is then defuzzified to form a crisp value PR. The above idea is implemented as follows:

associative (EX $\Rightarrow$ A, PR) $\leftarrow$

context (EX, CT),

contribute (CT, A, PD),

membership_function (priority, MF),

defuzzify (PD, MF, PR).

Essentially, releasing a component and sharing a component can be treated similarly, using the same decision-making procedure, but with different rule bases, as given in Figures 3 and 4. The low-level fuzzy logic operations are also implemented in Prolog.

■ CASE STUDY

An automated fish processing plant is used as an example to illustrate how the restructuring system works (de Silva and Gu, 1994). The initial operating condition (or normal operating point) is illustrated in phase I of Figure 8. There are three workcells in this system — Cutting Workcell, Grading Workcell, and Packaging Workcell — which have three types of sharable component — vision stations, robots, and AGVs (automated guided vehicles). In the initial configuration the components are assigned to workcells as follows:

- Cutting Workcell: Vision1, Vision4, Robot1, AGV2, AGV5;
- Grading Workcell: Vision2, Robot2, AGV3, AGV4;
- Packaging Workcell: Vision3, Robot3, Robot4, AGV1, AGV6.

Initially in phase I the component loads are assigned according to the process demand (task demands) by the load planner. The demand levels are assigned to be 100% in this stage of normal operation, as shown in Figure 8.

Updating Task Demand

Any change to the demands causes the KS of "Load Planning" to trigger. This KS will be executed in the next control cycle. It will then change the blackboards of "Workcell Information" and "Component

INTELLIGENT RESTRUCTURING OF PRODUCTION SYSTEMS 291

Information", which will trigger the KSs of "Sharing Feasibility Updating" and "Restructuring". Due to the preset priorities, the KS of "Sharing Feasibility Updating" will be fired first to modify the BB of "Sharing Feasibility". After all the presettings are made, the FPS will be restructured according to the information in the blackboard. The KS of "Restructuring" may change blackboards of "Workcell Information" and "Component Information", and also may post restructuring commands on the blackboard. The change in workcell information may trigger some KSs to confirm that the load has been properly distributed and to check the restructuring feasibility. The change of restructuring commands will trigger the output KS. Finally, the system will become idle again, for steady operation of the FPS.

As an example, suppose that the cutting demand drops by 40% due to a reduction in the raw material supply (e.g., at the end of a fishing season), and the grading demand is increased in 50% in order to reduce an existing backlog. Only the fish of high grade are packaged, and the low-grade fish may be used for canning. Therefore, suppose that according to the current market demand, the packaging load is maintained at the previous level. The conditions due to these changes in task demand are shown as "Phase II Transition" in Figure 8. Clearly, Robot2, AGV3, and AGV4 will be overloaded, resulting in a capacity shortage. Also, Vision1, Vision4, AGV2, and AGV5 will now operate well below their full capacity. The component status is updated as follows:

   component (robot 4, workstatus, [(lu, 1)]).
   component (vision1, workstatus, [(mu, 0.8), (hu, 0.6)]).
   component (vision4, workstatus, [(mu, 0.8), (hu, 0.6)]).
   component (robot1, workstatus, [(lu, 0.8), (mu, 1)]).
   component (agv2, workstatus, [(mu, 0.8), (hu, 0.6)]).
   component (agv5, workstatus, [(mu, 0.8), (hu, 0.6)]).
   component (vision3, workstatus, [(ok, 1)]).
   component (robot3, workstatus, [(ok, 1)]).
   component (agv1, workstatus, [(ok, 1)]).
   component (agv6, workstatus, [(ok, 1)]).
   component (vision2, workstatus, [(ok, 1)]).
   component (robot2, workstatus, [(lo, 1)]).
   component (agv3, workstatus, [(mo, 0.75), (lo, 0.25)]).
   component (agv4, workstatus, [(mo, 0.75), (lo, 0.25)]).

Sharing feasibility is checked for pairs of component in a group of sharable components. Here, factors such as component reliability, the geographical position of the components, and operating cost will be considered. The following feasibility matrix is used here:

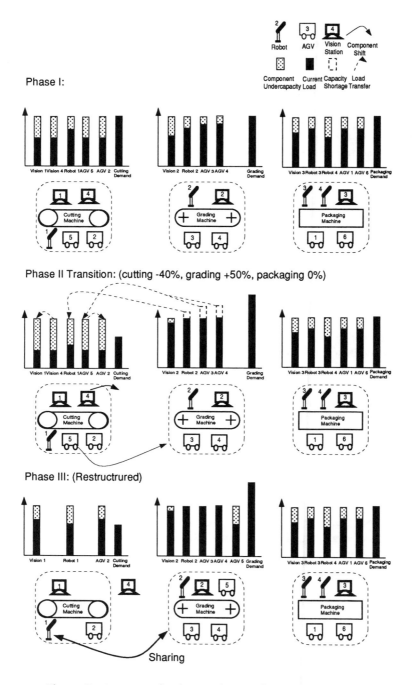

Figure 8 A case study: the simulation of a demand change.

INTELLIGENT RESTRUCTURING OF PRODUCTION SYSTEMS 293

	robot1	robot2	robot3	robot4
robot1	0.00	0.92	0.90	0.81
robot2	0.92	0.00	0.95	0.85
robot3	0.90	0.95	0.00	0.90
robot4	0.81	0.85	0.90	0.00

Now the KS of "Restructuring" is triggered. Note that the feasibility matrix should be updated at the end of every action. The following are some intermediate results of the simulation, which show the decision-making procedure:

1. Component releasing action.

FIND all undercapacity components of vision: [vision1,vision2,vision3,vision4]
To find all possible actions for vision1 within workcells.
action: releasing(vision1,vision4)
  fuzzy priority = [(vh,0.6),(hi,0.6),(md,0.6),(lw,0.8)]
  priority = 0.538
  feasibility = 0.9
  assessment = 0.48
To find all possible actions for vision2 within workcells.
To find all possible actions for vision3 within workcells.
To find all possible actions for vision4 within workcells.
action: releasing(vision4,vision1)
  fuzzy priority = [(vh,0.6),(hi,0.6),(md,0.6),(lw,0.8)]
  priority = 0.538
  feasibility = 0.9
  assessment = 0.48
The best action is releasing(vision4,vision1); assessment = 0.48
This action can be executed!
Load (transferred) for action releasing(vision4,vision1) is: 36.0
*** execute action:releasing(vision4,vision1)

2. Component sharing action.

FIND all overloaded components of robot: [robot2]
To find all possible actions for robot2 between workcells.
action: sharing(robot2,robot1)
  fuzzy priority = [(md,1.0)]
  priority = 0.4
  feasibility = 0.92
  assessment = 0.368
action: sharing(robot2,robot4)
  fuzzy priority = [(md,1.0)]
  priority = 0.4
  feasibility = 0.855
  assessment = 0.342
The best action is sharing (robot2,robot1); assessment = 0.368
This action can be executed!
Load (transferred) for action sharing(robot2,robot1) is: 20.0
*** execute action:sharing(robot2,robot1)

Similar procedures are used for other actions. The procedure will continue until none of the actions can be executed (i.e., when the assessment values fall below the threshold value). Then the planner will check the blackboard, calculate the cost function, and report the restructuring result:

> PLANNING RESULT
> plan:releasing(vision4,vision1)
> plan:releasing(agv5,agv2)
> plan:sharing(robot2,robot1)
> plan:shifting(agv5,from(cutting),to(grading))
> plan:sharing(agv4,agv5)
> plan:sharing(agv3,agv5)
> Restructuring succeeds; The cost function is: J=0.729.

The final result of restructured system is illustrated as phase III in Figure 8.

Feasibility Change

Another example is used to show how the feasibility values can affect the restructuring results. The same three-workcell fish processing system and same initial (normal) operating conditions as in the previous part of the case study are used here. This time, however, suppose that Robot1 is partially damaged (e.g., its network communication part is operating at a reduced speed). Consequently, its shareability is considerably reduced as shown in the feasibility matrix below:

	robot1	robot2	robot3	robot4
robot1	0.00	0.28	0.27	0.24
robot2	0.28	0.00	0.95	0.85
robot3	0.27	0.95	0.00	0.90
robot4	0.24	0.85	0.90	0.00

The feasibility change is considered when the planner selects a robot for sharing the load of Robot2. The decision-making procedure during the computer simulation is outlined below:

> FIND all overloaded components of robot: [robot2]
> To find all possible actions for robot2 between workcells.
> action: sharing(robot2,robot1)
>   fuzzy priority = [(md,1.0)]
>   priority = 0.4
>   feasibility = 0.276
>   assessment = 0.11
> action: sharing(robot2,robot4)
>   fuzzy priority = [(md,1.0)]
>   priority = 0.4
>   feasibility = 0.855
>   assessment = 0.342
> The best action is sharing(robot2,robot4);

```
         assessment = 0.342
      This action can be executed!
      Load (transferred) for action sharing(robot2,robot4) is: 20.0
      *** execute action:sharing(robot2,robot4)
```

The final result is shown in Figure 9, and the output is shown below:

```
      PLANNING RESULT
      plan:releasing(vision4,vision1)
      plan:releasing(agv5,agv2)
      plan:sharing(robot2,robot4)
      plan:shifting(agv5,from(cutting),to(grading))
      plan:sharing(agv4,agv5)
      plan:sharing(agv3,agv5)
      Restructuring Succeeds; The cost function is: J = 0.777.
```

The higher value of the cost function for this second case is justifiable in view of the degraded performance of a workcell component. Also, note that the final outcome (phase III of Figure 9) is somewhat different from what was realized in the previous case (phase III of Figure 8) for the same starting and transition conditions. This is a direct result of the change in feasibility parameters of component sharing.

■ SUMMARY

An appropriate criterion for optimal operation of a FPS is to make the system components operate at their designed capacity levels. However, in practice it is virtually impossible to exactly satisfy this requirement. Incompatibilities and disparities among system components, variations in the process demands and operating conditions, and the qualitative and incomplete nature of system information are reasons for this difficulty. In this chapter a knowledge-based approach was developed for automatic load sharing among components in a multiple-workcell FPS. In the approach the component associations with workcells could be modified for the purpose of load transfer, and consequently the system structure might be changed as well during process operation. Hence, this approach of load distribution is also known as "dynamic structuring". Fuzzy logic was used to represent and process knowledge that is associated with the examination of system components for the purpose of load distribution. The two main load transfer processes considered were (1) absorbing by a component that operates below its capacity, the excess load of an overloaded component of the same type, and (2) absorbing the load from an undercapacity component by another undercapacity component of the same type, so that the first component could be released. The qualitative, experience-based, and incomplete nature of the load transfer knowledge made the use of fuzzy logic particularly

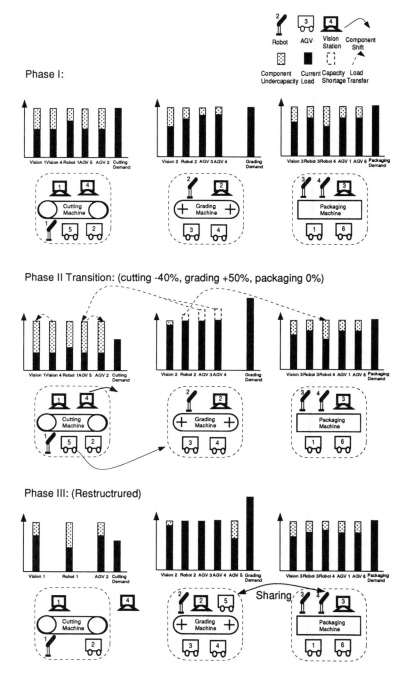

Figure 9 A case study: the simulation of a sharing feasibility change.

INTELLIGENT RESTRUCTURING OF PRODUCTION SYSTEMS 297

suitable in the present problem. The feasibility of load transfer between two components was expressed by a *feasibility matrix* which took into account such factors as the ease of sharing (proximity, compatibility, etc.), operating conditions of the components (wear and tear, repair state, etc.), and the operating cost per unit processing rate of a task. Load transfer priorities were determined by the knowledge-based system of FPS restructuring through fuzzy logic in conjunction with the feasibility indices. A *blackboard architecture* was developed for implementation of the knowledge-based restructuring system, using *Prolog*. A case study of a three-workcell, automated, fish processing plant was described to illustrate the application of the approach developed in this chapter.

■ PROBLEMS

1. What are advantages and drawbacks of distributed intelligence over central intelligence, in the context of a multi-component production system? In particular, do groups of objects perform more intelligently than individual objects with very little interaction?

2. Consider a supervisory intelligent system of a flexible manufacturing system. In particular, consider the *Task-Sequence Planner*, which is a subsystem of the supervisory system. Its tasks include assignment of various subtasks to active agents, coordination of the operation of the agents, supervision of the operation of these agents, and rescheduling in case of overloading, degradation, or malfunction. Indicate various needs of information/knowledge processing within this subsystem, and state whether fuzzy logic is suitable in each such need.

3. Hierarchical architecture, blackboard architecture, and object-oriented architecture are three possibilities of implementing the *knowledge system* of an intelligent controller. In fact, *blackboards* or *frames* and *objects* may be organized into hierarchical architectures and hence, the latter two architectures are not mutually exclusive from the first. In a three-level hierarchy, the top layer may carry out non-real time activities such as planning and scheduling while the intermediate layer may perform coordination or time-synchronization, and the bottom layer may perform low-level process executions in real time. Outline advantages of a hierarchical structure of this type in comparison with a non-hierarchical blackboard or object-oriented implementation.

4. Consider a flexible production system (FPS) of the type considered in Chapter 9, with an optimal scheme for dynamic restructuring. Inventory level may be used as a criterion for optimality. Explain why "zero inventory" may not be a realistic objective, and rather "minimum inventory" would be preferred. Discuss how a fuzzy, rule-based approach would be consistent with this preference.

   In the quadratic cost function given by Equation (6), explain how the numerical values would be assigned to the weighting parameters α_c.

5. Sensors may be part of a workcell component, and may be termed "component sensors". They may be used for various purposes such as low-level control and high-level monitoring. Alternatively, sensors may be considered in the context of monitoring workcell (or component) tasks. Here, the term "task sensor" is used. Identify the differences between component sensors and task sensors.

6. Development of a flexible production system will involve three hierarchical stages:
   1. Sensor and controller integration for various components and machine tools (Lowest level).
   2. Integration of each workcell using the components in Stage 1. Inter-component communication, that is compatible with an open architecture, so that non-proprietary components may be added, would be a consideration here (Intermediate level).
   3. Plant (facility) automation. Here, linking various workcells, and developing the upper-level management systems will be involved (Top level).

   Flexibility is desirable, particularly when production in small batches (e.g., parts-on-demand) is needed. Note that reconfiguration and retooling of a non-flexible system can be quite costly and time consuming. Discuss how a flexible production system may cope with an unfamiliar or unexpected condition or disturbance input.

7. Figure 10 shows schematic representations of two stages of a flexible production system (de Silva, 1993a,b). Identify the category of the system and indicate when and why such a system would be advantageous. How could this system benefit from a knowledge-based (particularly, fuzzy logic) approach?

INTELLIGENT RESTRUCTURING OF PRODUCTION SYSTEMS 299

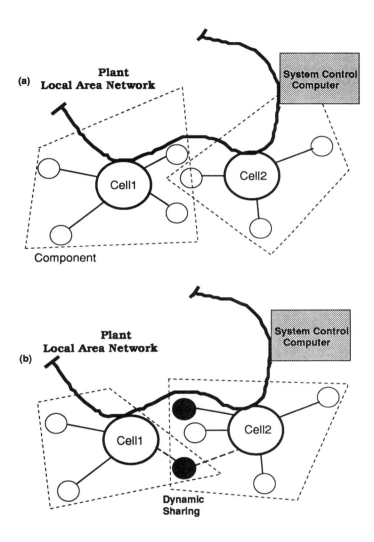

Figure 10 Two stages of component configuration of a flexible production system.

■ REFERENCES

de Silva, C.W. (1989). *Control Sensors and Actuators,* Prentice-Hall, Englewood Cliffs, NJ.

de Silva, C.W. (1991). Fuzzy Information and Degree of Resolution Within the Context of a Control Hierarchy, *Proc. IEEE Int. Conf. Industrial Electronics, Control, and Instrumentation,* Kobe, Japan, pp. 1590–1595.

de Silva, C.W. (1993a). Soft Automation of Industrial Processes, *Eng. Appl. Artif. Intelligence,* Vol. 6, No. 2, pp. 87–90.

de Silva, C.W. (1993b). Knowledge-Based Dynamic Structuring of Process Control Systems, *Proc. 5th Int. Fuzzy Systems Assoc. World Congress,* Seoul, Korea, Vol. 2, pp. 1137–1140.

de Silva, C.W. and MacFarlane, A.G.J. (1989). *Knowledge-Based Control with Application to Robots,* Springer-Verlag, Berlin.

de Silva, C.W. and Gu, J. (1994). An Intelligent System for Dynamic Sharing of Workcell Components in Process Automation, *Eng. Appl. Artif. Intelligence Int. J.,* Vol. 7, No. 5, pp. 571–586.

Dubois, D. and Prade, H. (1980). *Fuzzy Sets and Systems,* Academic Press, Orlando, FL.

Engelmore, R.S. and Morgan, T. (Eds.). (1988). *Blackboard Systems,* Addison-Wesley, Reading, MA.

Jaganathan, V., Dodhiawala, R., and Banna, L.S. (Eds.). (1989) *Blackboard Architectures and Applications,* Academic Press, San Diego, CA.

Kovacs, G. and Mezgar, I. (1991). Expert Systems for Manufacturing Cell Simulation and Design, *Eng. Appl. Artif. Intelligence,* Vol. 4, No. 6, pp. 417–424.

Weatherhall, A. (1978). *Computer Integrated Manufacturing,* Butterworths, London.

Zadeh, L.A. (1979). Theory of Approximate Reasoning, in *Machine Intelligence,* Hayes, J. et al. (Eds.), Vol. 9, pp. 149–194.

10 FUTURE APPLICATIONS

INTRODUCTION

Fuzzy logic control is not a panacea, but it can play an important role in making automated processes more intelligent. Due to various complexities in an industrial process, mathematical modeling can become quite difficult, but a sufficient knowledge base might be available through past experience and knowledge concerning the process. If this knowledge base can be expressed as a set of linguistic protocols that contain fuzzy descriptors, there lies a quite favorable environment for applying fuzzy logic. When a functional hierarchy exists within the automated process, fuzzy logic control is particularly suitable for implementation at the upper levels, rather than for low-level direct control.

In addition to high level control, fuzzy logic can be useful in intelligent multiagent control, autonomous robotics, intelligent sensing and actuation, and in the emerging discipline of *mechatronics*. This concluding chapter investigates several applications of fuzzy logic in the context of intelligent or soft automation. The topics addressed here are somewhat futuristic, and applied research is being carried out on these various subjects at present (de Silva, 1994).

INTELLIGENCE IN AUTOMATION

What is termed "hard" or "fixed" automation deals with industrial processes which are fixed and repetitive in nature. Here, the system configuration and the operations are fixed and cannot be varied without incurring considerable down time and cost. It has served us well, however, over many decades, particularly in applications that call for fast and accurate operation, when manufacturing large batches of the same product. As retooling and extensive resetting would be needed for changing the product in a plant that employs hard automation, more flexible and adaptable architectures of automation have been developed. In *flexible automation* the task is programmable and a workcell may be quickly reconfigured to accommodate a change in product. It is particularly suitable for plant environments in which a variety of products are manufactured in small batches. This has obvious advantages, even though a trade-off usually exists between the speed and flexibility of operation. Automated processes, particularly in the realm of flexible automation, may encounter unexpected or previously unknown conditions. Some degree of *intelligence* would be required in handling them. Furthermore, fully or partially *autonomous* operation, self-reconfiguration, diagnosis, and self-tuning would need high-level, knowledge-based decision making. The associated knowledge base may contain qualitative, incomplete, approximate, and noncrisp (fuzzy) aspects, and may rely on experience-based learning rather than a detailed analysis of physical principles, or analytical modeling. Here we enter the area of "soft automation" (de Silva, 1993a), which involves flexible and intelligent operation of an automated process. The subject of *automation intelligence* primarily deals with the technology of incorporating "intelligence" into an automated process. The representation and processing of knowledge plays a key role in implementing intelligence for process automation. Fuzzy logic has emerged as a leading approach for accomplishing this, and is particularly appropriate when a process is complex, incompletely known, slow, and difficult to model either analytically or experimentally. The main requirement, however, is that a knowledge base should be available, that may be expressed as a set of linguistic protocols which contain fuzzy descriptors such as "slight", "fast", and "accurate", for carrying out the task.

Sensing and *sensor fusion* are important in knowledge acquisition, and directly in process monitoring for both low-level feedback control and high-level *supervisory control*. The control architecture must be designed to suit the application. For example, distributed multiagent systems with reactive control will benefit from localized intelligence in each agent, distributed throughout the system. Hierarchical architectures are appropriate in processes that possess functional hierarchies. Typically, more intelligence is needed in decision making at higher levels in the hierarchy. *Mechatronics* is an emerging discipline in which mechanics and electronics are combined through a concurrent and integrated approach of design and development in the manufacture of *smart products*. As demonstrated by this concluding chapter, all these areas, which play important roles in automation intelligence, can benefit from the application of fuzzy logic. However, the control applications of fuzzy logic are not limited

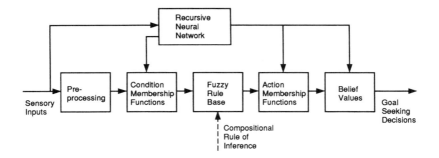

Figure 1 A fuzzy-neural decision-making system for goal seeking.

to those that directly fall within process automation, as we have seen in the previous chapters of this book.

INTELLIGENT MULTIAGENT CONTROL

In process applications we come across situations of distributed control in which a multitude of machines or devices work toward a specific objective. Examples include parts transfer using groups of mobile robots on a factory floor, and production tasks that involve multiple machine tools within a flexible manufacturing system. The individual process tools (agents) may work cooperatively where the tasks and task loads may be redistributed by an upper level supervisory controller. Alternatively, the agents may operate quite separately in a competitive and *reactive* manner. In either situation, the performance within the distributed control environment can significantly improve by incorporating a degree of intelligence into the agents themselves, with or without a higher-level intelligent controller for supervisory actions. When the agents function in a competitive and reactive manner, a process of *natural evolution* may occur, with the elimination of weaker or undesirable agents in some form of "natural selection". Furthermore, the intelligence that is incorporated in a process environment of this type should be able to accurately reflect characteristics and capabilities such as "learning through mistakes", and the commonly experienced superiority of group actions over independent individual actions. The nature of the partnership of agents must be properly represented by giving due consideration to the level of the partnership.

Consider the decision-making problem associated with goal seeking by multiple agents in a process environment. A fuzzy-neural architecture, as schematically represented in Figure 1, may be appropriate here. It contains a knowledge base which is expressed in terms of fuzzy if-then rules that represent the logical *goal-seeking* actions for various conditions of the process. There is a "belief value" associated with each such rule, which represents the *level of belief* by *domain experts* of the validity of that rule. In addition to the rules themselves, the membership functions of the condition and action variables must be known. The belief values and the parameters of the membership functions are initially determined off line using a static, *feedforward neural network* employing a training set of data obtained from the process, with input from domain experts. These parameters may be updated on line during operation of the goal-seeking system, using a *dynamic* or *recurrent neural network* (Rao and Gupta, 1993). The output of the knowledge-based system will be a goal-seeking decision corresponding to a particular context, as determined by the sensory inputs from the process and its environment.

The rule base may be established using fundamental knowledge of the process and its environment, process objectives, past experience, and expert opinion. In using fundamental knowledge the decision-making problem may be formulated as one of optimizing a *cost function*. The cost function itself expresses the cost of making a particular goal-seeking decision under a given set of process conditions and for a specified set of process objectives. The cost weightings may be assigned through experience and expertise. Once the knowledge base, the membership functions, and the belief values for the rules are all known, the on-line decision making may be carried out by applying the compositional rule of inference (Zadeh, 1979) as described in Chapters 3 and 4. The steps involved would be

1. Fuzzify the sensory information. Some preprocessing might be needed here to make the low-level sensory information compatible with the high-level knowledge base (de Silva, 1991a). Details are found in Chapter 8.
2. Supply the fuzzified sensory information to the knowledge base, and apply the compositional rule of inference.
3. Interpret the goal-seeking decision obtained in step 2, and execute it.

■ RECONFIGURABLE AUTONOMOUS MANIPULATORS

The use of *autonomous* robot systems has tremendous advantages, particularly in space applications, with regard to accuracy, repeatability, flexibility, speed, and minimal intervention of crew members (Smith et al., 1989; de Silva, 1991b). Such complex systems, however, must possess reliable and effective control systems. Most advanced techniques of robotic control are applied at the lowest direct-control level, as seen in the earlier chapters of the book.

There are disadvantages to applying complex control techniques, and particularly those that require an analytical or experimental model of the process at the lowest level of a control system. First, it may not be feasible to obtain a sufficiently accurate analytical model. Second, on-line model identification and application of complex control algorithms can result in degraded speed of performance. Also, associated delays can lead to large errors and instability (Bialkowski, 1986). In an autonomous robotic system, then, it is also necessary to have a high-level *supervisory controller* that is able to not only monitor and "tune" the performance of the low-level direct controllers, but also will have the necessary "intelligence" for controlling the robot operation in the presence of incomplete information, uncertainty, and unplanned or unfamiliar conditions. A hierarchical control architecture, as described in Chapters 6 and 8, may be suitable for robotic manipulators in order to provide some degree of intelligence for autonomous operation (de Silva and MacFarlane, 1989), and may be applicable in adaptive, reconfigurable manipulator systems.

A three-level hierarchy may be used in the control system. Here, the lowest layer will consist of direct, algorithmic control of the conventional type, and associated sensors and actuators. The intermediate layer will be an *intelligent preprocessor* which will monitor robot information from the high-resolution sensors at the bottom layer and will interpret that information for use in the top layer. A knowledge-based *supervisory* and *self-organization* module will occupy the top layer of the hierarchy. At this level the context information established by the intelligent preprocessor will be used in the rule base and processed by applying a suitable rule of inference in order to arrive at high-level control decisions. The rule base is established using expert knowledge, available experience, heuristics, and any other information (including theoretical concepts and computer simulations) that might be available. In view of the incompleteness of this knowledge and possible qualitative and uncertain nature of the information, fuzzy logic would be an appropriate means for both representation and processing of the knowledge. As before, a belief value may be incorporated into each fuzzy logic rule in the high-level knowledge base. An integrated *learning scheme* may be incorporated as well into the robotic control system.

■ INTELLIGENT FUSION OF SENSORS AND ACTUATORS

In multiple-sensor fusion several sensors acquire the same information, and thereby increase the reliability and accuracy of the information through a systematic process of combination of the sensory data obtained from different sources (Luo and Kay, 1989). Typically, statistical approaches like the *Bayesian estimation* are used for sensor fusion, in view of the uncertainty associated with the information from different sensors. Furthermore, because the knowledge used in sensor fusion may be heuristic, some work has been carried out in applying knowledge-based approaches in that context (Pau, 1989). Known applications of sensor fusion include robotics and flexible manufacturing (Thien and Hill, 1991) and process monitoring (Alexander et al., 1989). The focus of the present section is the development of an integrated approach for the "fusion" of multiple sensors and actuators so as to reach the performance of distributed sensors and actuators. This technology of *spatially distributed* (continuous) sensors and actuators will directly benefit the automation of industrial processes, particularly in relation to instrumentation cost, speed, and accuracy of operation. Accurate and complete information regarding a process can make monitoring and control of the process quite effective. Increasing the number and complexity of the process sensors, however, will increase the instrumentation cost and also can decrease the control bandwidth and system reliability (de Silva, 1989). Similarly, the accuracy of mechanical processing, including object manipulation, can be improved by increasing the number of actuators, but the same disadvantages as for sensors would apply. It follows that reducing the number and the complexity of the sensors and actuators, as proposed here, can bring about direct benefits to an industrial process, by making the implementation simpler, more reliable, and cost effective.

In theory, distributed sensors and actuators are desirable for industrial processes, but in practice, often it is not cost effective and sometimes it is not altogether feasible to use distributed sensors and actuators for process applications. In view of this dichotomy, consider an integrated approach for implementing a finite set of discrete sensors and actuators to yield a performance that is "close" to what might be achieved from spatially distributed and continuous sensors and actuators. A control system for this purpose may have a layered architecture, as schematically shown in Figure 2.

The discrete sensors and actuators will occupy the lowest layer of the system. The intermediate layer will have information preprocessors (filters) for the sensors along with dynamic models of the process and its environment. The uppermost layer will consist of a knowledge-based decision-making system. This intelligent module will examine the preprocessed sensory data, interpret them as distributed information, and will facilitate the operation of the industrial process by providing high-level control commands for the actuators. The preprocessed sensory information will be used for monitoring the performance of the overall process. Furthermore,

FUTURE APPLICATIONS 307

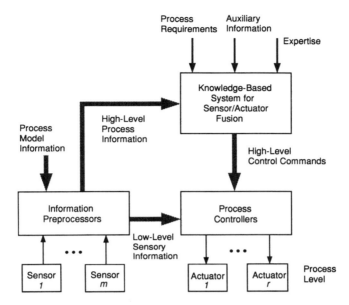

Figure 2 Basic structure of a hierarchical system for sensor/actuator fusion.

low-resolution, high-level information obtained using the preprocessors will form the "context" for driving the top-level knowledge-based system, as discussed in Chapter 8 (de Silva, 1993b). The top layer of the architecture will use the preprocessed sensory information about the process and its environment, compare it to the desired performance (i.e., for distributed sensors and actuators), using the knowledge base, and will make decisions as to how the individual, discrete actuators should be controlled in order to reach the desired level of performance. This level may be implemented as an object-oriented, knowledge-based system. Various forms of knowledge, including that of human experts, what is "learned" over a period of time using sensory information, basic analytical concepts, and even common sense and heuristics would be useful in this layer. A *blackboard-type, object-oriented* representation would be suitable here. The knowledge base may be incomplete, uncertain, and qualitative, at least in part. Consequently, the use of fuzzy-neural approaches in representing the knowledge and in implementing the reasoning and decision-making process, will be appropriate.

The present problem may pose some analytical considerations as well. First, a model for information preprocessing (de Silva, 1993b) may be required for extracting sensory information in a compatible manner for the top-level knowledge-based system. In particular, the necessary *transitional* and *combinational* operations must be developed, as stated in Chapter 8. Second, the sensor/actuator fusion that

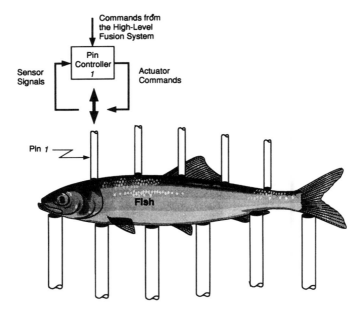

Figure 3 A prototype system for robotic processing of fish.

is addressed here may be considered an optimization problem. Specifically, a *performance index* may be established that would represent the cost of deviating from the performance of an idealized situation of distributed sensors and actuators, and a parameter optimization may be carried out. This could be used, for example, to determine such information as the locations and various parameter values of the discrete sensors and actuators that would minimize the cost function. Third, analytical models of the process and its environment might be needed for performance comparison, particularly in physical implementation. Some analytical study will be useful also in dynamic modeling, and in integrating such models into the optimization procedure.

As a physical implantation of the present approach, consider a flexible robotic device for handling and processing fish, as schematically shown in Figure 3. It consists of a three-dimensional matrix of programmable pins. Each pin may be controlled independently with a local controller, and associated with it there are sensors (e.g., for force and motion sensing). The layered control system, as outlined above, is linked with the direct controllers of the manipulator pins.

■ MECHATRONICS ERA

Mechatronics is a new discipline which concerns the integrated design, manufacture, and control of smart electromechanical products

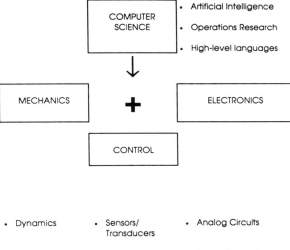

Figure 4 The discipline of mechatronics.

(de Silva, 1994). In this marriage of *mechanics* and *electronics* a unified approach is used for handling the mechanical and electronic aspects of the design. Here, mechanical aspects are not limited to structural or solid-mechanics considerations, but also cover fluid mechanics and thermodynamics. The operation of an integrated electromechanical device of this nature will be facilitated by intelligent (smart, knowledge based) control, with appropriate sensors, actuators, digital and analog electronics, and signal conversion and communication circuitry. A schematic representation of the discipline of mechatronics is given in Figure 4.

The integration and unification of electrical and mechanical elements may be accomplished by treating the overall system as a set of interacting subsystems that are connected through *energy-transfer ports*. Conversion between various forms of energy (e.g., kinetic, potential, thermal, electromagnetic, electrostatic) may occur through these ports. By using appropriate analogies for basic processes and lumped elements in the system (e.g., force-current analogy) in representing energy storage, dissipation, and input sources, a mixed-mode approach would be feasible in the unified design and analysis procedure.

At the planning and supervisory levels of the system, operations-research and artificial intelligence considerations may be utilized. Knowledge representation and processing may employ fuzzy logic in these upper layers, as outlined before. Conventional, "crisp control" should be exploited as much as possible for low-level control. In view of the integrated and unified approach used in mechatronics, the overall design should be more realistic, efficient, optimal, reliable, and cost effective. Also, due to integrated intelligence, the *machine intelligence quotient* (MIQ) should be high. Undoubtedly, mechatronics is an area, as is robotics, in which a synergy should be established among engineers from the fields of computer science, electrical engineering, and mechanical engineering toward a common, noble objective of developing smarter products. A related field of *chematronics* may be developed with the involvement of chemical engineers.

■ CONCLUSION

To some extent, fuzzy logic has been treated unfairly by the control systems community. This has manifested in two extremes. First, there are those who reject fuzzy logic control as an avocation and outrightly refuse to study and understand the contributions made in this field, merely by emphasizing the historic negative connotations of the term "fuzzy". There are others who commit an "overkill" by attempting to apply fuzzy logic to every conceivable control problem, thereby inviting a largely unjust criticism that the peer review process to which the technical papers in this area are subjected is less rigorous. It is our responsibility to take appropriate countermeasures in dealing with such criticism.

Fuzzy logic control is not a panacea. As pointed out throughout this book, there are specific situations in which the application of fuzzy logic control is quite appropriate and even preferred. We in the control engineering community should recognize these situations and get involved. The present book is a step toward this objective.

You have noticed the "applications" flavor of the book, even though theoretical developments are equally important. Some sections of the book have dealt entirely with fuzzy logic considerations. In some others, fuzzy logic has been employed as just one of several tools used in achieving a specific goal. Theory has been integrated as much as possible throughout the book, and an attempt has been made to present it in a simplified manner, with suitable graphic illustrations.

PROBLEMS

1. In a conventional approach to developing an electromechanical system, the mechanical components are typically decided upon first, and then the electronics and controls are added on as needed by the mechanical system. Controllers used are relatively simple, and operating precision is achieved through fine tolerances and stiff construction, which would lead to costly manufacturing. Contrast this with the *mechatronics* approach.

2. In conventional ("hard") computing, imprecision, fuzziness, and uncertainty are considered undesirable attributes. But in "soft computing" approximate reasoning approaches such as fuzzy logic are exploited where the approximate nature of the problem and the associated reasoning scheme are not necessarily negative characteristics. Outline a problem of intelligent control where soft computing would be more appropriate than hard computing.

3. In sensor fusion, multiple sensors are used to monitor a given feature. By combining the information from different sensors, the overall accuracy of the measured information can be improved and the sensory efficiency can be increased. Consider the fusion of *vision* and *tactile sensing*. It is known that some *dexterous* tasks may be carried out more accurately and at high speed by experienced workers, using tactile sensing associated with human "touch". However, when vision is also incorporated, the speed and sometimes accuracy have been known to decrease in some such tasks. As an example consider the buttoning of a collar. In some other applications (e.g., production and process operations), vision and tactile sensing have been used in "fusion", with better performance. Indicate advantages and shortcomings of sensor fusion. Could sensor fusion be considered in the category of intelligent sensors?

4. It is known that a serious weakness of a computer in making knowledge-based decision is its lack of common sense. It makes syntactic arguments rather than semantic ones. Also, it uses one "best" scheme at a time, rather than a balanced "soft" approach of using several strategies in parallel. Extend such arguments towards a case for *fuzzy-logic computers* (or "soft computers").

5. Define the term "mechatronics". Give important considerations which have to be addressed when developing a control system for a mechatronic device.

REFERENCES

Alexander, S.M., Vaidya, C.M., and Kamel, K.A. (1989). An Architecture for Sensor Fusion in Intelligent Process Monitoring, *Comput. Ind. Eng.*, Vol. 16(2), pp. 307–311.

Bialkowski, W.L. (1986). Control Systems Engineering: Have We Gone Wrong?, *InTech*, Vol. 2, pp. 27–32.

de Silva, C.W. (1989). *Control Sensors and Actuators*, Prentice-Hall, Englewood Cliffs, NJ.

de Silva, C.W. (1991a). Fuzzy Information and Degree of Resolution Within the Context of a Control Hierarchy, *Proc. IEEE Conf. Industrial Electronics, Control, and Instrumentation*, Kobe, Japan, pp. 1590–1595.

de Silva, C.W. (1991b). Trajectory Design for Robotic Manipulators in Space Applications, *J. Guidance, Control, Dyn.*, Vol. 14(3), pp. 670–674.

de Silva, C.W. (1993a). Soft Automation of Industrial Processes, *Eng. Appl. Artif. Intelligence*, Vol. 6(2), pp. 87–90.

de Silva, C.W. (1993b). Hierarchical Processing of Information in Fuzzy Logic Control Applications, *Proc. 2nd IEEE Conf. Control Applications*, Vancouver, Canada, Vol. 1, pp. 457–461.

de Silva, C.W. (1994). Automation Intelligence, *Eng. Appl. Artif. Intelligence, Int. J.*, Vol. 7(5), pp. 471–477.

de Silva, C.W. and MacFarlane, A.G.J. (1989). *Knowledge-Based Control with Application to Robots*, Springer-Verlag, Berlin.

Luo, R.C. and Kay, M.C. (1989). Multi-Sensor Integration and Fusion in Intelligent Systems, *IEEE Trans. Syst. Man Cybernetics*, Vol. 19(5), pp. 901–931.

Pau, L. (1989). Knowledge Representation Approaches in Sensor Fusion, *Automatica*, Vol. 25(2), pp. 207–214.

Rao, D.H. and Gupta, M.M. (1993). A Multi-Functional Dynamic Neural Processor for Control Applications, *Proc. 1993 Am. Control Conf.*, San Francisco, pp. 2902–2906.

Smith, J.H., Estus, J., Heneghan, C., and Nainan, C. (1989). The Space Station Freedom Evolution Phase: Crew-EVA Demand for Robotic Substitution by Task Primitives, *Proc. IEEE Int. Conf. Robotics and Automation*, Scottsdale, AZ, pp. 1472–1477.

Thien, R.J. and Hill, S.D. (1991). Sensor Fusion for Automated Assembly Using an Expert System Shell, *Proc. 5th Int. Conf. Advanced Robotics*, Pisa, Italy, Vol. 2, pp. 1270–1274.

Zadeh, L. A. (1979). Theory of Approximate Reasoning, *Machine Intelligence*, Hayes, J. et al. (Eds.), Vol. 9, pp. 149–194.

APPENDIX A

FURTHER TOPICS ON FUZZY LOGIC

■ INTRODUCTION

In this book we have attempted to avoid complex mathematical analysis as much as possible. In doing so, we have limited our discussion of several analytical concepts. In this appendix, we shall revisit some of these topics.

■ GENERALIZED COMPOSITION

In Zadeh's compositional rule of inference, the two operations *min* and *max* are applied on membership functions (Zadeh, 1973). It is understood that *min* is used when selecting the least value among a set of discrete membership functions or grades, and "infimum" (or *inf*) is used when the least value over a continuous interval is needed. Similarly, the *max* operation is used when the largest value among a set of discrete membership functions or grades is determined, and the "supremum" (or *sup*) is used instead, when the largest value over a continuous interval is needed. All these operations are considered binary in the sense that each operation may be applied to *two* membership functions (or grades) at a time, resulting in one membership function (or grade). The associated mapping is indicated by (Dubois and Prade, 1980):

$$f : [0,1] \times [0,1] \to [0,1] \qquad (A.1)$$

This fact should be clear because each one of the two membership functions that are operated upon (operands or arguments) takes values in the *real-unit interval* [0, 1] and the membership function that results from this operation also takes values in the unit interval. The operation itself is denoted by "f" in Equation (A.1) and may denote any one of *min*, *inf*, *max*, and *sup*. In view of the triangular nature of the mapping given by Equation (A.1) the operation is termed a *triangular norm*. These ideas can be generalized as outlined below.

Triangular Norm (T-Norm)
As listed in Table A.1, a triangular normal (or T-norm) is a binary operation or function of the type given by Equation (A.1) that possesses the following four properties:

1. It is nondecreasing in each argument.
2. It satisfies commutativity.
3. It satisfies associativity.
4. It satisfies the boundary condition $xT1 = x$.

It can be shown that a second boundary condition

$$xT0 = 0$$

is also satisfied. To prove this, set $x = 0$ in Property 4. Hence, $0T1 = 0$. However, from Property 2, we have $1T0 = 0$. Now, in view of Property 1, it follows that, $1T0 \geq xT0$. Since $xT0$ is in the real interval [0, 1] it cannot be less than 0. Hence, $xT0 = 0$.

Three examples of the T-norm are given in Table A.1. Note that the *min* operation is a special case of a T-norm. Hence we may interpret the T-norm as a *generalized intersection* that may be applied to fuzzy sets (or membership functions).

TABLE A.1

SOME PROPERTIES OF A TRIANGULAR NORM		
Item description	**T-Norm (triangular norm)**	**S-Norm (T-conorm)**
Function	$T : [0, 1] \times [0, 1] \to [0, 1]$	Same
Nondecreasing in each argument	If $y \geq x, z \geq w$ then $yTz \geq xTw$	Same
Commutative	$xTy = yTx$	Same
Associative	$(xTy)Tz = xT(yTz)$	Same
Boundary conditions	$xT1 = x$ $xT0 = 0$	$xS0 = x$ $xS1 = 1$
	with $x, y, z, w \in [0, 1]$	
Examples	$min(x, y)$ xy $max[0, x + y - 1]$	$max(x, y)$ $x + y - xy$ $min[1, x + y]$
De Morgan law	$xSy = 1 - (1 - x) \, T(1 - y)$	

Triangular Conorm (S-Norm)
The complementary operation of a T-norm is called an S-norm. As indicated in Table A.1, this is also a binary operation of the form given by Equation (A.1) which possesses the following four properties:

1. It is nondecreasing in each argument.
2. It satisfies commutativity.
3. It satisfies associativity.
4. It satisfies the boundary condition, $xS0 = x$.

In parallel with the T-Norm, it can be shown that a second boundary condition, $xS1 = 1$, is also automatically satisfied.

Exercise: Using the four properties of a S-Norm, show that the boundary condition $xS1 = 1$ holds.

Three examples of an S-Norm are given in Table A.1. Note that the *max* operation is a special case of S-Norm and that this norm may be interpreted as a *generalized union*. The theory of fuzzy logic may be generalized by extending the composition to the general case that uses the S-norm and the T-norm rather than *max* and *min* operations. De Morgan law, as given in Table A.1, is also satisfied by these two generalized norms.

FUZZY CONTROLLER REQUIREMENTS

In fuzzy-logic inference, and particularly in fuzzy-logic control, one may run into difficulties unless certain conditions are satisfied by the associated knowledge base. Specifically, the following conditions should be met:

1. The rule base should be "complete".
2. The rules should not "interact".
3. The rules should be "consistent".
4. The inferences should be "continuous".
5. The rule-based system should be "robust" and "stable".

It should be stated that these are somewhat ideal requirements, some of which being contradictory and some others complementary. Now, let us indicate the meaning of each of these requirements.

Specifically, consider a fuzzy-logic rule base R with the condition fuzzy descriptors (i.e., context or antecedent fuzzy sets) $\mathbf{X}$ and the action fuzzy descriptors (i.e., consequent fuzzy sets) $\mathbf{C}$, such that,

$$R \cup_i X_i \to C_i \tag{A.2}$$

Now, as discussed in Chapters 3 and 4, for a particular context (or response measurement or observation) $\hat{X}_j$ the corresponding fuzzy logic inference $\hat{C}_j$ is determined by applying the *composition operation*, as

$$\hat{X}_j \circ R = \hat{C}_j \tag{A.3}$$

This notation is used in the following discussion.

Completeness

A rule base is said to be complete if it provides a meaningful inference (control action) for any possible context (system response). In other words, at least one rule has to be specified for every possible condition of the system. More specifically, for any possible condition $\hat{X}$ in the observation space **X** there should be at least one corresponding rule i $(X_i \rightarrow C_i)$ such that the membership function of the consequent C_i is not null;

$$\forall_{\hat{X} \in \mathbf{X}} \exists X_i \rightarrow C_i \quad \text{such that } hgt(C_i) > 0 \tag{A.4}$$

Here, $hgt(C_i)$ is the *height of the membership function* of C_i, as defined by

$$hgt(C_i) = \sup_c \mu_{C_i}(c) \tag{A.5}$$

which is simply the global peak value of the membership function.

Rule base completeness need not be strictly satisfied for a rule-based system to operate satisfactorily. Often, less important or unknown rules are skipped from a rule base without seriously affecting the system performance. In the matrix representation of a rule base, the empty grid elements are an indication of an incomplete rule base.

Rule Interaction

Consider the rule base R that is given by Equation (A.2). We say that the rules in the rule base interact, if there exists ($\exists$) some rule i such that

$$X_i \circ R \neq C_i \tag{A.6}$$

In other words, if the rules in the rule base interact, then the *antecedent* (context) X_i of a particular rule, when composed with the rule base R, may not necessarily result in the *consequent* C_i as given by the rule itself (Pedrycz, 1989).

Some care should be exercised in interpreting this condition. If the rule base R is formed by just one rule, then if the antecedent X of this rule is composed with R the result will be C, which is the consequent of the original rule itself. Of course, this assumes that consistent logical operations are used both in the formation of R and in the application of the compositional rule (for example, *min* operation for implication and *sup-min* for composition). This should be intuitively clear because there cannot be rule interaction unless there is more than one rule. When there are several rules, however, the stronger rules will tend to dominate, and rule interaction may be present. Then the condition (context) of a weak rule, when composed with the overall rule base, may not provide exactly the action (consequent) as predicated by the rule itself. It should be clear that such interaction is not necessarily a bad thing because a weaker rule would be an uncertain or inaccurate rule.

Consistency

A rule base is said to be consistent when there are no contradictory rules. Suppose that a rule base has the following two rules:

$$X_i \rightarrow C_i$$
$$X_i \rightarrow C'_i \qquad (A.7)$$

Such that for the same context (antecedent) X_i two consequents C_i and C'_i are possible. If these consequents are mutually exclusive; i.e.,

$$C_i \cap C'_i = \phi \qquad (A.8)$$

then, the rule base is said to be *inconsistent*. If a rule base R is inconsistent, the fuzzy-logic inference that is obtained through the composition

$$\hat{C} = \hat{X} \circ R \qquad (A.9)$$

can be *multimodal*. That means, the membership function $\mu_{\hat{C}}(c)$ of the inference can have more than one peak. This is likely the case when there are two or more contradictory rules that are very strong.

Continuity

A rule base is said to be continuous if, when the condition fuzzy sets overlap, the action fuzzy sets also overlap. To analytically express this condition, first let us define the *possibility function* of two fuzzy sets A and B as the membership function of the set intersection $A \cap B$. Specifically, the possibility function

$$F_{A,B}^{(x)} = \mu_{A \cap B}^{(x)} \qquad (A.10)$$

with $x \in X$ where X is the universe of discourse of both A and B.

Then, a rule base is continuous if and only if for all rule pairs i and j:

$$X_i \rightarrow C_i$$
$$X_j \rightarrow C_j$$

with non-null condition possibility (i.e., $X_i \cap X_j \neq \phi$) the action possibility will also be non-null (i.e., $C_i \cap C_j \neq \phi$).

Continuity in a control rule base can help provide smooth and chatter-free performance in the control system. As in the case of *centroid defuzzification*, a continuous rule base may result in a slow (sluggish) performance, however.

Robustness and Stability

Robustness of a control system denotes the insensitivity to parameter changes, model errors (in model-based control), and disturbances (including disturbance inputs and noise in various control signals). Stability of a control system refers to bounded response when the inputs themselves are bounded. This is termed *bounded-input-bounded-output* (BIBO) stability. A special case is the *asymptotic stability* where the response asymptotically approaches zero (origin) when excited from the origin and the inputs are maintained at zero value thereafter. Then, robustness may be interpreted as stability under system disturbances.

The degree of stability of a control system is a measure of distance to the state of *marginal stability* (i.e., almost unstable or steadily oscillating state under zero input conditions). This distance is termed the *stability margin* of which *phase margin* and *gain margin* being special measures for linear systems in the frequency domain (see, for example, de Silva, 1989). Similarly, a *robustness index* of a control system may be determined by establishing a representative bound for a system parameter or signal disturbance within which the control system will remain stable but outside which it is likely to become unstable. Note, however, that fuzzy-logic control is not a model-based technique in the sense that it does not employ an explicit model of the process. Consequently, in this case, robustness cannot be defined with respect to model errors. Since robustness is traditionally defined in terms of stability, we shall explore the latter topic further.

■ STABILITY OF FUZZY SYSTEMS

The subject of stability of a fuzzy system (particularly, of a fuzzy-logic control system) has been introduced in Chapter 5. As noted, one of two approaches may be used to study stability in a fuzzy system:

1. Represent the fuzzy controller by a nonlinear model, through simplifying assumptions, and perform stability analysis using the traditional approaches.
2. Interpret stability as that of the fuzzy-logic inference mechanism and study stability of the decision-making system, largely at the system level.

The first approach is the one that is predominantly used in the published literature, perhaps due to the convenience and availability of well-established techniques, particularly Lyapunov-like approaches for the stability analysis of nonlinear systems (see, for example, Safonov, 1980). However, this is a somewhat artificial (or synthetic) way of dealing with stability of a fuzzy system, and may not be generally valid. The reason is simple. Once the fuzzy subsystem is represented by an analytic nonlinear model with nonfuzzy parameters and variables, one no longer has a fuzzy system with its inherent fuzzy features. Specifically, consider the nonlinear state space model given by

$$\frac{d\mathbf{x}}{dt} = \mathbf{f}(\mathbf{x}, \mathbf{u}) \qquad (A.11)$$

$$\mathbf{y} = \mathbf{g}(\mathbf{x}, \mathbf{u}) \qquad (A.12)$$

where $\mathbf{x} \in \Re^n$ is the state vector, $\mathbf{y} \in \Re^m$ is the output (system response) vector, and $\mathbf{u} \in \Re^r$ is the input (control action) vector.

Suppose that a fuzzy-logic rule base R is available that relates the fuzzy response descriptors $\mathbf{Y}$ to fuzzy control action descriptors $\mathbf{U}$:

$$R : \mathbf{Y} \to \mathbf{U} \qquad (A.13)$$

A crisp response $\mathbf{y}$ has to be first fuzzified using a FZ operator (say, reading off the membership grades from the membership functions of $\mathbf{Y}$):

$$\hat{\mathbf{Y}} = FZ(\mathbf{y}) \qquad (A.14)$$

Next, apply the compositional rule of inference to obtain a fuzzy control action $\hat{\mathbf{U}}$;

$$\hat{\mathbf{U}} = \hat{\mathbf{Y}} \circ R \qquad (A.15)$$

and finally, defuzzify using the FZ^{-1} operator (say, employing the *centroid method*) as:

$$\mathbf{u} = FZ^{-1}\hat{\mathbf{U}} \qquad (A.16)$$

The overall nonlinear controller that is equivalent to the fuzzy controller is then,

$$\mathbf{u} = \mathbf{h}(\mathbf{y}) \qquad (A.17)$$

where,

$$\mathbf{h} = FZ^{-1}(FZ(\mathbf{y}) \circ R) \qquad (A.18)$$

Note that even though the nonlinear controller given by Equation (A.17) looks quite innocent and may be employed in the traditional techniques of stability analysis, it is in fact an intractable function as given by Equation (A.18), unless greatly simplifying assumptions with regard to, for instance, vector dimensions, rules, and the membership functions, are made.

The second approach to the stability analysis of fuzzy systems concerns the stability of the fuzzy decision maker itself. To illustrate this approach, consider a scalar fuzzy system f (i.e., a state equation) with fixed input u, and expressed in the discrete-time form:

$$X_{n+1} = f(X_n, u) \qquad (A.19)$$

where, X_i is considered a fuzzy variable of system state. The system response is given by applying the compositional rule of inference (say, *sup-min*):

$$X_{n+1} = X_n \circ R \qquad (A.20)$$

or

$$\mu_{X_{n+1}}(x_{n+1}) = \sup_{x_n \in X} \min\{\mu_{X_n}(x_n), \mu_F(x_{n+1}, x_n)\} \qquad (A.21)$$

where R denotes the fuzzy-logic rule base which relates X_i to X_{i+1}:

$$R : X_i \rightarrow X_{i+1} \qquad (A.22)$$

Then, through successive application of the state transition relation, we get

$$X_{n+1} = X_1 \circ R^n \qquad (A.23)$$

where $R^n = R$ composed n-1 times.

Now, the fuzzy system is *stable* if $\lim_{n \to \infty} R^n$ exists, without oscillations. Note, however, that this approach is also somewhat philosophical and further analysis is possible only for special cases.

■ CONCEPTS OF APPROXIMATION

Fuzziness may be considered as a concept of approximation, and it plays an important role in *approximate reasoning*. There are many

TABLE A.2
SOME CONCEPTS OF APPROXIMATION

Concept	Character	Example
Ambiguity	Has two or more (finite number of) possibilities.	The value of x may or may not be 3.0.
Generality	Has a variable that can take any possible value.	The value is x.
Imprecision	Can assume a value within a finite tolerance (interval of crisp values).	The value of x lies between 1.0 and 4.0.
Uncertainty or probability	There is a probability associated with the event.	There is a 75% probability that the value of x is 3.0.
Vagueness	Does not have meaningfully defined limits.	The value of x may be about 3.0.
Fuzziness	The membership of a set is not sharply (crisply) defined.	The value of x is small.
Belief (subjective probability)	The degree of belief (through knowledge and evidence) of the membership of a set (crisp or fuzzy).	The statement "3 is an element of set A" is believed at a level of 80%.
Plausibility	The dual of belief. If Belief $(x \in A) = b$, then Plausibility $(x \notin A) = 1 - b$.	The statement "3 is an element of set A" is plausible at a level of 90%.

other concepts of approximation that are not quite synonymous with fuzziness, but are often used interchangeably, albeit inexactly. Several such concepts are listed in Table A.2. In particular, note that *fuzziness* is represented by a "fuzzy set" which does not have a crisp boundary. As a result, its membership is not clearly defined. For example, the fuzzy descriptor "small" is a fuzzy set. Whether a particular value (say, 3.0) belongs to this set is a matter of perception and is often subjective. Hence, a grade of membership within the real interval [0,1] may be assigned to the set of elements in the universe of discourse, depending on the level of perception that a particular element is a member of the set.

As should be clear from Table A.2, *vagueness*, in particular, is not the same as *fuzziness*, even though often used to denote it. In fact, vagueness is much more general than fuzziness. Another misconception is that uncertainty and associated *randomness* is identical to fuzziness. It should be emphasized that randomness denotes the *uncertainty* of the membership of an object in a *crisp* (non-fuzzy) set. Hence, one may associate a degree of uncertainty (or, *probability*) that a particular object belongs to a (crisp) set. However, once the uncertainty is resolved, there remains no randomness, and either the object certainly belongs to the set or it does not. Consequently, it should be clear that randomness is quite different from fuzziness. This distinction is further illustrated in Table A.3. In particular,

TABLE A.3
FUZZINESS AND RANDOMNESS

Concept	Randomness (uncertainty)	Fuzziness
Representation	Probability function	Possibility function
Explanation of symbolic representation	$Pr[x \in A] = a$ Probability of x belonging to a crisp set A is a. *Note:* The boundary of set A is fixed.	$\mu_A(x) = a$ Membership grade of element x in a fuzzy set A is a. *Note:* The boundary of set A is not sharply defined.
Character	Statistical inexactness due to random events.	Inexactness due to the human perception process.
Example	Age of x is 20 years or more.	Beauty of x.

randomness is represented by a *probability function* whereas fuzziness is represented by a *possibility* (or *membership*) function.

The confusion with regard to various concepts of approximation arises due to the similarity of their numerical/analytical representations. In particular, the concepts that are listed in Table A.2 may be analytically represented by similar measures (Dubois and Prade, 1980; Klir and Folger, 1988). Once these measures are quantified, then their distinctive characteristics are lost and could lead to incorrect perceptions of similarity.

Belief and Plausibility

Belief and plausibility are also concepts of approximation, and may be defined regardless of whether or not a set is fuzzy. Consider a set A which may or may not be fuzzy, in universe X. Then the statement,

$$\text{"Element } x \text{ belongs to set } A\text{"} \quad (A.24)$$

may be assigned a "belief value" $Bl(A)$ in the real interval [0, 1] which represents the *degree of belief* of the statement (A.24). The value assigned in this manner may be *subjective* and may depend on various factors such as available knowledge, experimentation and other evidence. In particular,

$$Bl(X) = 1; Bl(\phi) = 0 \quad (A.25)$$

where ϕ is the null set (or $\overline{X}$, the complement of the universe X).

For two sets A and B, both in the same universe X, the following axiom holds:

$$Bl(A \cup B) \geq Bl(A) + Bl(B) - Bl(A \cap B) \quad (A.26)$$

which may be generalized to the case of more than two sets. It follows from Equations (A.26) and (A.25) that,

$$Bl(A) + Bl(\overline{A}) \leq 1 \qquad (A.27)$$

where $\overline{A}$ is the complement of A. Specifically, Equation (A.27) states that $Bl(\overline{A})$ is not the mutual dual of $Bl(A)$. In particular, a weak belief of the statement (A.24) does not necessarily indicate a strong belief that x is outside the set A. This should be intuitively clear because a lack of knowledge or evidence of the factuality of the statement (A.24) does not imply that it is false.

Note: In case of probabilities, however, we have

$$Pr(A) + Pr(\overline{A}) = 1 \qquad (A.28)$$

This may be interpreted as a stronger belief, and consequently, *probability* is sometimes termed *Bayesian belief*, and the ordinary belief is sometimes termed *subjective probability*. In particular, from the statistical point of view of probability, the validity of Equation (A.28) should be clear because if we conduct an experiment and if the outcome belongs to A, then it certainly does not belong to $\overline{A}$ and vice-versa. Then, after a large (infinite) number of experiments we establish Equation (A.28).

Now, returning to Equation (A.27), the shortcoming that is implied may be removed by introducing the concept of *plausibility*. Specifically, the plausibility of A, denoted as $Pl(A)$, is defined to satisfy the relation:

$$Bl(A) + Pl(\overline{A}) = 1 \qquad (A.29)$$

In other words, the "belief of x belonging to A" is the *mutual dual* of the "plausibility that x does not belong to A". It is clear from the Inequality (A.27) and Equation (A.29) that

$$Pl(\overline{A}) \geq Bl(\overline{A})$$

and hence,

$$Pl(A) \geq Bl(A) \qquad (A.30)$$

and evidently, plausibility is a more general (or more trustworthy or believable) form of belief. Plausibility satisfies the following conditions:

$$Pl(X) = 1; Pl(\phi) = 0 \qquad (A.31)$$

and

$$Pl(A \cap B) \leq Pl(A) + Pl(B) - Pl(A \cup B) \qquad (A.32)$$

As in the case of belief, Inequality (A.32) may be generalized to more than two sets.

Rule Validity

A rule in a knowledge base becomes a fuzzy statement when the *rule-antecedent* (context) and the *rule-consequent* (action) are fuzzy descriptors represented by fuzzy sets. Apart from the fuzziness of these two types of descriptors, the validity of the rule itself may be questioned. A level of validity may be assigned to the rule depending on such factors as the amount of past experience, evidence, and overall knowledge about the rule. The level of validity may be modified as more knowledge is gained through *learning* and other means.

Consider the fuzzy-logic rule,

$$\text{IF } A \text{ THEN } B \tag{A.33}$$

The decision based on this rule may be affected by several factors. For example,

1. If the perception of either of the fuzzy descriptors A and B changes, then the associated membership function (μ_A or μ_B) will change and as a result the decision based on the rule may change.
2. If the data (observations, measurements, etc.) that determine the context of A change as a result of a change in state of the fuzzy-logic system, the corresponding grade of membership of the context will change (in the fuzzification operation) and consequently the decision based on the rule may change even when the perception of the fuzzy descriptors (membership functions of) A and B have not changed.
3. If through learning and further knowledge the level of belief of the rule itself changes, then the decision based on the rule may change even though the membership functions of A and B and the context data for A have not changed.

The consideration of Item (3) above may be incorporated into the rule (A.33) by introducing a *validity factor* b as:

$$\text{IF } A \text{ THEN } B \text{ WITH Belief } b \tag{A.34}$$

where $b \in ([0, 1]$.

Here, the processing of the rule may proceed as with (A.33), but the resulting fuzzy-inference (membership function) has to be scaled by b prior to rule combination and defuzzification. The result will be identical to scaling by b the membership function of B, but the underlying concepts (the membership function of B and the belief level of the rule) are quite different. Furthermore, the parameter b in rule (A.34) may represent any compatible concept such as "confidence", "certainty", "belief", "plausibility", "validity", and "feasibility" of the rule. This idea is used with regard to *feasibility index* in the application described in Chapter 9.

■ REFERENCES

de Silva, C.W. (1989). *Control Sensors and Actuators*, Prentice-Hall, Englewood Cliffs, NJ.

Dubois, D. and Prade, H. (1980). *Fuzzy Sets and Systems*, Academic Press, Orlando, FL.

Klir, G.J. and Folger, T.A. (1988). *Fuzzy Sets, Uncertainty and Information*, Prentice Hall, Englewood Cliffs, NJ.

Pedrycz, W. (1989). *Fuzzy Control and Fuzzy Systems*, Research Studies Press, Ltd., Somerset, England.

Safonov, M.G. (1980). *Stability and Robustness of Multivariable Feedback Systems*, MIT Press, Cambridge, MA.

Zadeh, L.A. (1973). Outline of a New Approach to the Analysis of Complex Systems and Decision Processes, *IEEE Trans. Systems, Man and Cybernetics*, Vol. 3, No. 1, pp. 28–44.

APPENDIX B: SOFTWARE TOOLS FOR FUZZY LOGIC APPLICATIONS

INTRODUCTION

Several software tools are available for developing fuzzy logic applications which may be subsequently implemented either in software or hardware. These tools range from high-level programming languages such as PROLOG which are not specific to fuzzy logic, to various environments and development systems for fuzzy logic implementations. The MUSE toolkit that was described in Chapter 6 is a general development system of this type, for rule-based artificial-intelligence applications. The main steps of programming a fuzzy logic application (including computer simulations) are

1. Specification of the rules (fuzzy knowledge base).
2. Specifications of the membership functions for the rule variables.
3. Definition of links between the input data (e.g., process responses) and the rule context of the fuzzy rule base; and the links between the rule inferences of the fuzzy system and the actions (e.g., controller inputs) to the external system.

In this appendix, we will provide an outline of a few typical tools or approaches for developing fuzzy-logic applications. This is not an exhaustive study and is intended for providing a flavor for what is feasible.

PROLOG

At the lowest level, a fuzzy logic application may be programmed using a high-level, universal programming language such as C. Since such languages are quite general and not specifically developed for knowledge-based applications, not to mention fuzzy logic, this will neither be efficient nor convenient. It is preferable, then, to use a high-level language such as PROLOG that is intended for knowledge representation and processing purposes.

The term PROLOG stands for PROgramming in LOGic (Clocksin and Mellish, 1981) and is a language that can program and interpret rules, in the form of *Horn clauses*. In a Horn clause there is exactly one none-negated literal, with the implication arrow reversed. For example, we may have

$$C \leftarrow A, B \qquad (B.1)$$

which represents *"A∩ B then C"*, or in the linguistic rule form: "If *A* and *B* then *C*". Here, *C* is the *literal*.

Backward chaining (see Chapter 2) is used in rule processing with PROLOG. The main step here is "unification" where the *PROLOG interpreter* attempts substitutions so as to make the "goal literal" equal to the "trial literal". In a PROLOG program, this procedure is carried out in the order in which the clauses occur within the program. Hence, it is necessary to program the clauses in the order of their complexity (simple ones first), in order to avoid endless or lengthy searches.

The following syntax of PROLOG should be noted:

1. Negation is denoted by "¬"
2. Intersection ("OR") is denoted by a comma ","
3. Union ("AND") is denoted by a semicolon ";"
4. A clause must end with a period "."
5. A variable must begin with an upper-case letter
6. A constant or a predicate name must begin with a lower-case letter

The reasoning process begins when a "query" is issued. The "unification" takes place, by trying all substitutions one after the other until either the goal is achieved or the search is failed.

Example

Consider the PROLOG data:

workcell (*cell-1*, *rob-2*).
system (*sys-fp*, *cell-1*).

and the rule:

syst-comp (*A,C*) ← *system* (*A,X*), *workcell* (*X,C*).

Here, the data indicate that the robot *rob-2* is a component of the workcell *cell-1*, which in turn belongs to the fish processing system *syst-fp*. Suppose that the query

? - *workcell* (*What*, *rob-2*)

is issued. This intends to determine the workcell to which the robot *rob-2* belongs. Then *"What"* is unified with *"cell-1"* in the first item of data, providing the goal *What* = *cell-1*. Next, if the query

$$?\text{-}syst\text{-}comp(syst\text{-}fp, What)$$

is issued, the PROLOG rule as given before, will be used. First A is instantiated to *syst-fp*, in unifying the right-hand side (predicate) of the rule, from left to right. Then X will be unified to *cell-1* in view of the second item of data. Finally, C will be unified with *rob-2*, in view of the first item of data. Hence, the goal of *What* = *rob-2* will be achieved. Note that it is quite possible to have other items of compatible data. Then, the very first match is considered the final goal. For example, if the data items were:

workcell (*cell-1, rob-1*).
workcell (*cell-1, rob-2*).
system (*syst-fp, cell-1*).

the goal returned by the PROLOG program would be *What* = *rob-1* rather than *What* = *rob-2*, because the former is the first match, even though both answers are correct.

A case study of applying PROLOG to a high-level process control application that employs fuzzy logic is described in Chapter 9. Further details are found in (de Silva and Gu, 1994; Gu, 1994). It should be noted here that even though PROLOG has an efficient interpretation scheme, it has limitations. In particular, more than one non-negated literal is not allowed in a rule.

■ FUZZY-SYSTEM DEVELOPMENT ENVIRONMENT

A convenient software environment for developing fuzzy-logic based applications, both as software and hardware, is commercially available from Togai Infralogic (Togai, 1991). High-level and research-based applications where speed of operation is not a serious limitation but flexibility (convenience in modification and expansion) is important, software implementations are preferred. In commercial applications such as consumer appliances (e.g., video cameras, air conditioners, television sets, toasters, washing machines, vacuum cleaners) where controllers are fixed, mass-produced at low cost, and have to be fast and compact, hardware implementations that employ "fuzzy chips" are the norm.

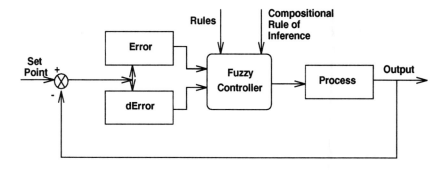

Figure B1 A direct in-loop fuzzy logic control system.

Fuzzy Programming Language

The fuzzy programming language (FPL) of Togai Infralogic (TIL) is a special-purpose, high-level language that is used with fuzzy-logic based expert-system tools. The language uses several template objects such as *PROJECT, FUZZY, RULE, VAR, MEMBER*, and *CONNECT*, to define a fuzzy-logic problem. This is done manually, using the graphical environment *TILShell* in association with the *Fuzzy-C Development System*. A "C" program file of the problem is generated in this manner, using the *Fuzzy-C Compiler*. Finally, this C code may be *compiled*, and *linked* with other program-object codes to generate the overall *executable code* of the application.

Programming a Fuzzy-Logic Control Problem

Consider a typical, in-loop fuzzy-logic control problem as shown by the block diagram in Figure 1. Here, the response error (*Error*) and the change in error (*dError*) are used as the context variables for the fuzzy-logic rules. The output of the fuzzy-logic controller is the defuzzified control action that drives the process. The main steps involved in the generation of this controller using *TIL Fuzzy-C Development system* are given below:

1. Define the System Structure in the TILShell using the *Project Editor* as shown in Figure 2. Here the variable objects (*VAR*) *Error, dError* and *Control* and the fuzzy rule base object (*FUZZY*) *Control-Rule* are generated and connected, satisfying the appropriate causality (from-to) using *CONNECT* objects.
2. Define the variables of the project using the *Variable Editor*. As shown in Figure 3, one should specify the following. Whether the variable is crisp or fuzzy; variable storage type; universe of discourse; initial value; default output (if it is an output variable that is not set by the rule base); and support set and data type for the membership function (if fuzzy).
3. Create membership functions for the fuzzy variables of the application, using the *Membership Function Editor.* As shown in Figure 4,

SOFTWARE TOOLS FOR FUZZY LOGIC APPLICATIONS 331

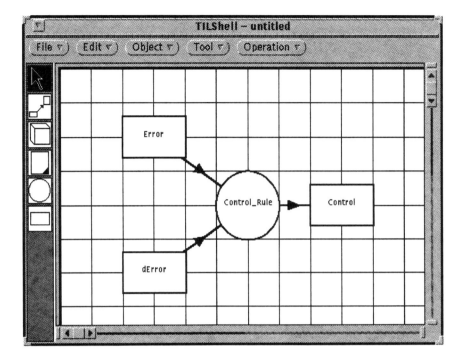

Figure B2 Definition of the project structure using TIL shell.

one can create and name a membership function either as a collection of points or a mathematical equation, in this manner.
4. Create the fuzzy-logic rules of the control rule base, using the *Fuzzy Editor*. As shown in Figure 5, one may generate rules and *fragment objects* (objects which contain C code procedures), edit, delete, or view the rules and fragment objects using this editor.
5. Specify the particulars of each rule in the rule base, using the *Rule Editor*. As shown in Figure 6, one may create and modify the *condition* (antecedent) and *action* (consequent) parts of a rule, add or delete rules, and loop through rules.

If one is not quite confident about manually specifying the rule base, a tool called *TILGen* may be used. This tool uses input and output data that are specified by the user (say, as obtained from experimental observation or specified as the desired performance by a control engineer), and automatically generates a set of fuzzy rules using a neural network, thereby greatly simplifying the task of developing a fuzzy-logic expert system. However, the validity of the rule base generated in this manner cannot be fully guaranteed, and directly depends on the associated neural-network procedures.

Once an application is generated as outlined above, the associated *fuzzy programming language* code is compiled into a C code

332 INTELLIGENT CONTROL: FUZZY LOGIC APPLICATIONS

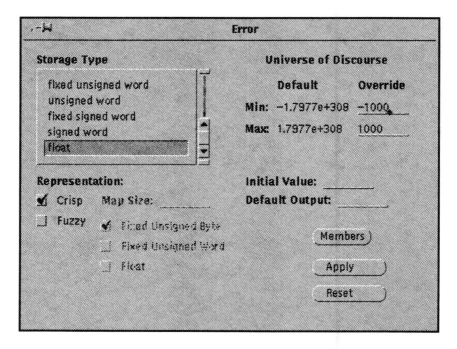

Figure B3 Specifying a variable using the VAR editor.

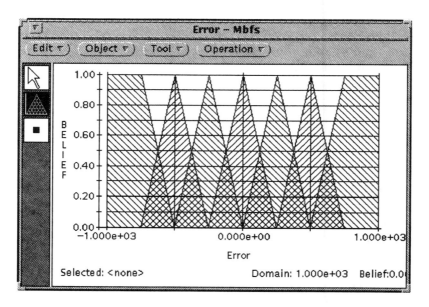

Figure B4 Creating the membership function of a fuzzy variable using the membership function editor.

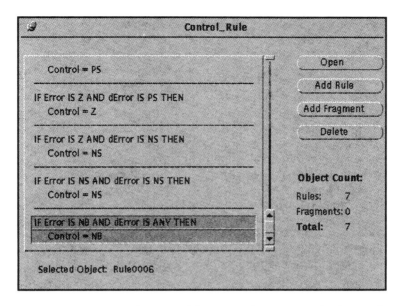

Figure B5 Generation of rules and fragment objects using the fuzzy editor.

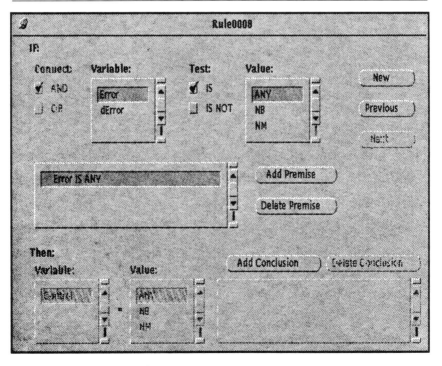

Figure B6 Editing a rule using the rule editor.

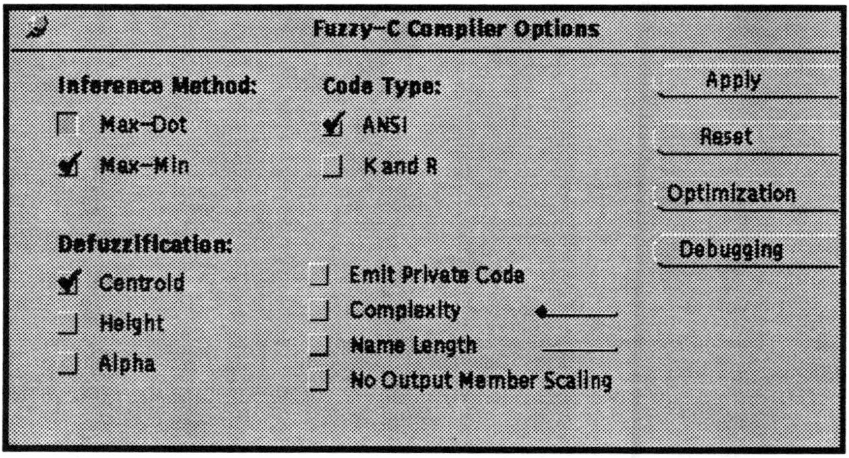

Figure B7 Compiling an application into a C code using the fuzzy-C compiler.

using the Fuzzy-C compiler. In doing so, as shown in Figure 7, one may choose a composition (inference) method (e.g., *max-dot* or *max-min*), a defuzzification method, and the standard of the C code (e.g., ANSI or the older "K and R" standard).

Hardware Implementation

Once an application is developed using TILShell, as outlined in the previous section, it may be utilized in its software form to control a process, employing say, a personal computer as the controller. However, for large-quantity applications such as in consumer appliances and electronics, where the controller is fixed but where low cost, small size and fast speed are crucial requirements, the hardware implementation through fuzzy chips is irreplaceable.

A single-board fuzzy controller (Jamshidi, 1993) may be developed, for example, using the FC110 Development System of Togai Infralogic. The Compiler of FC110 first converts the fuzzy programming language (FPL) code generated by TILShell into an object file of machine code of the specific microprocessor that is used as the controller. There is also the main program (or the *driver code*) which coordinates various input-output activities of the process interface. This is also converted into machine code by the FC110 Assembler. The resulting machine-code object files are linked by the FC110 Linker, and is available as the *executable code* for the hardware controller. The board will contain it's own analog to digital conversion (ADC) and digital to analog conversion (DAC) chips, a BUS and other communication hardware.

REFERENCES

Clocksin, W. and Mellish, C. (1981). *Programming in PROLOG*, Springer, New York.

de Silva, C.W. and Gu, J. (1994). An Intelligent System for Dynamic Sharing of Workcell Components in Process Automation, *Eng. Appl. Artific. Intel. Int. J.*, Vol. 7(5), pp. 571–586.

Gu, J. (1994). Knowledge-Based Dynamic Restructuring of Flexible Production Systems, M.A.Sc. thesis, Department of Mechanical Engineering, The University of British Columbia, Vancouver, Canada.

Jamshidi, M. (1993). Fuzzy Logic Software and Hardware, *Fuzzy Logic and Control*, Jamshidi, M., Vadiee, N., and Ross, T.J. (Eds.), PTR Prentice Hall, Englewood Cliffs, NJ.

Togai Infralogic (1991). *Fuzzy-C Development System User's Manual*, Togai Infralogic, Irvine, CA.

INDEX

a posteriori conditional probability, 67
a priori probability, 67
Absorption, 28
Abstraction level, 264
Action lines, 246
Action, see Consequent
Action variables, 212, 217
Activity of component, 273
Adaptive control, 10
Alpha-cut (α-cut), 49, 123, 244
    threshold methods, 80
Ambiguity, 103, 321
Antecedent, 34, 107
Approximate reasoning, 320
Architectures of fuzzy control, 81
Artificial intelligence, 1, 24
Associative, 66, 289
Associativity, 28, 314
Asymmetry index, 261
Asymptotic stability, 318
Attributes, 171
Automation intelligence, 302
Automation, 301
Autonomous manipulators, 304
Autonomous operation, 148
Autonomous system, 302
Averaged guess method, 61

Backward chaining, 35, 41
    systems (BCS), 36
Bandwidth, 11, 34, 219
Bayes' relation (Bayes' theorem), 67
Bayesian belief, 323
Bayesian estimation, 306
Belief function, 98
Belief level, 304
Belief, 321
BIBO stability, 318
Binary polling, 61
Blackboard architecture, 197, 281

Blackboard representation, 307
Bold union, 108
Bottom-up approach, 35
Boundary conditions, 28, 66, 314
Bounded max, 108
Bounded min, 108
Bounded sum, 48
Brushless DC motor, 216

Cartesian product, 50, 52
Center of gravity method, see Centroid
    method
Central intelligence, 297
Centroid defuzzification, 318
Centroid equation, 226
Centroid method, 72, 99, 111, 182, 226
Combinational operation, 157, 247, 307
Comfort control system, 77, 94
Command generation, 253
Common sense, 69
Commutative, 66
Commutativity, 28, 314
Compensator, 5, 219
Complement (NOT), 25, 47
Completeness of rule base, 316
Component coordination, 253
Component sensor, 298
Composition, 54, 57
Compositional rule of inference, 58, 72, 76, 87, 110, 224
Computational efficiency, 117
Condition, see Antecedent
Condition variables, 217
Conditional probability, 67
Conflict resolution, 36, 39, 164, 288
Conjunction rule of inference, 28
Connectives, 25
Consequent, 35, 108
Consistency of rule base, 317
Context knowledge, 31

Continuity of rule base, 317
Control bandwidth, 74, 119, 238
Control hierarchy, 160
Control line, 39
Control module, 248
Control unit, 281, 285
Controller attributes, 211, 219
Cost function, 274, 295, 304
Coupled rule base, 224
Crisp algorithm, 147
Crisp controller, 147
Crisp measurements, 75, 114
Crossover frequency, 219
Cutting workcell, 290
Cylindrical extension, 56, 227

Damping ratio, 217
Damping, 11
Data averaging, 248
Data base, 34
Data-driven search, 35
Decision table for servo tuning, 183
Decision table, 83
Decomposition theorem, 125
Decoupling control, 162
Decoupling controller, 7
Decoupling of rule base, 224
Defuzzification, 71, 75, 78, 111
    centroid method, 79, 80
    mean of maxima method, 79, 80
    threshold methods, 80
Degree of belief, 322
Degree of fuzziness, 237, 243
Delta function, 75
DeMorgan law, 27, 28, 314
Derivative time constant, 6
Descriptive terms, 236
Detectability, 16
Digital signal processing (DSP), 129
Dirac delta function, 115
Direct control, 152
Direct digital control (DDC), 70
Dissociated formulation, 120
Distributed actuators, 306
Distributed control, 235

Distributed intelligence, 297
Distributed sensors, 306
Distributivity, 28
Divergence, 116, 132
Domain blackboard, 283
Domain experts, 304
Domain knowledge, 283
Dynamic coupling, 7
Dynamic interaction, 5
Dynamic sharing, 269
Dynamic switching, 122

Eigen-fuzzy sets, 67
Energy transfer port, 309
Exclusion, 28
Exclusive OR, 25
Expectational knowledge, 31
Experience, 103
Expert control, 104
Expert systems, 37, 96
Expertise, 24, 103
Extension principle, 52, 67, 165

Factual knowledge, 31
Feasibility index, 277
Feedback control, 4
Feedback linearization, 7
Feedback module, 247
Feedforward neural networks, 304
Filter module, 247
Fine manipulation, 205
First match, 36
Fish processing machine, 240
Fish processing plant, 290
Fixed automation, 301
Flexible automation, 302
Flexible manufacturing system (FMS), 264
Flexible production system (FPS), 269
Flexible systems, 2
FMS, see Flexible manufacturing systems
Forward chaining, 35
FPL, see Fuzzy programming language
FPS, see Flexible production systems
Frames, 31, 297
    action, 31

INDEX 339

schematic, 32
situational, 31
Freeway incident detection, 97
Functional module, 248
Fuzzification, 70, 74, 96
Fuzziness, 49, 123, 243, 249, 251, 265, 321
Fuzzy associative memory (FAM), 98, 119
Fuzzy chips, 334
Fuzzy control, 69
Fuzzy controller, 176
Fuzzy decision making, 288, 320
Fuzzy dynamic systems, 54
Fuzzy label, 44
Fuzzy logic, 43
Fuzzy programming language (FPL), 330
Fuzzy relations, 50
Fuzzy resolution, 98, 108, 114, 119, 122, 241
Fuzzy self-tuning, 130
Fuzzy sets, 44
Fuzzy singleton, 74, 110, 114, 226
Fuzzy tuner, 82
Fuzzy tuning, 112
Fuzzy-C, 330
Fuzzy-neural architecture, 303

Gain margin, 318
Generality, 103, 321
Goal-seeking actions, 303
Grading workcell, 290
Gross motion, 205

H_∞ (H-infinity) control, 16
H_∞ norm, 18
Hard automation, 301
Hardware implementation, 334
Height of membership function, 228
Heuristics of restructuring, 287
Heuristics, 69
Hierarchical architecture, 297
Hierarchical control structure, 151
Hierarchical control system, 156
Hierarchical fuzzy control, 235
Hierarchical model, 241
Hierarchical structure, 82, 239, 271
Hierarchical system, 254
High-level fuzzy control, 156

Idempotency, 28
Impedance control, 203
Implication (IF-THEN), 48
Implication, 26
Imprecision, 103
Incremental rule base, 125
Inference engine, 34
Inference mechanism, 23
Inference, 27, 55
Infimum (inf) operation, 313
Information prefilter, 259
Information preprocessing, 156
Information processing, 249
Information resolution, 241
Information, 24
   structured, 24
Injection molding, 96
Integral controller, 221
Integral time constant, 6
Intelligence, 24, 265, 302
Intelligent monitoring, 252
Intelligent preprocessor, 271, 305
Intelligent process control, 140
Intelligent restructuring, 266
Intelligent sensing, 272
Intelligent sensor, 158, 264
Intelligent system, 23
Intelligent tuning, 255
Interaction of rules, 110
Intermodal spacing, 242
Interpolative reasoning, 233
Intersection (AND), 26, 47
Intuitive relation, 61
Inverse kinematics problem, 160
Inverse kinematics, 175
Inverted pendulum, 92
Involution property, 65
Iron butcher, 2, 235

Join, 57

Kalman filter, 16
Knowledge base, 23, 34, 236
Knowledge engineer, 37
Knowledge engineering, 41, 96
Knowledge source (KS), 167, 281

Knowledge system, 271, 275, 297
Knowledge, 24, 31, 103, 265
    context, 31
    expectational, 31
    factual, 31
    procedural, 24
    processing, 23, 27
    representation, 23, 27
Knowledge-based restructuring, 275
Knowledge-based supervisor, 264
Knowledge-based tuning, 103, 213
KS, see Knowledge source

Law of exclusion, 43
Laws of logic, 27
Lead compensator, 216, 220, 221
Learning control, 10
Learning scheme, 305
Linear quadratic Gaussian control (LQG), 15
Linguistic rules, 69
Linguistic terms, 237
Linguistic tuning rules, 152
Logic processing, 27
Logic, 23
Low-level controller, 82
LQG, see Linear quadratic Gaussian control

Machine intelligence quotient (MIQ), 310
Machine intelligence, 24, 145
Magnitude of compensator, 219
Marginal stability, 318
Massaging robot, 204
Match, 36
    first, 36
    most recent, 36
    privileged, 36
    toughest, 36
max operation, 47, 71, 313
max-dot, 66
max-min composition, 100
max-min, 66
Mean of max method, 72, 99
Mechatronics, 308, 311

Membership function, 44, 132
    estimation of, 60
Membership grade, 44
min operation, 47, 53, 71, 313
MIQ, see Machine intelligence quotient
MIT rule, 11, 105
Mixed mode approach, 309
Model identification, 8
Model-based control, 2, 146
Model-referenced adaptive control, 10
Modus ponens rule of inference, 28
Monitoring, 158
Most recent match, 36
Multiagent control, 303
Multimodal membership, 79, 317
MUSE AI toolkit, 166, 327

Natural evolution, 303
Natural frequency, 217
    damped, 217
    undamped, 218
No-activity component, 276
Nonlinear feedback control, 7
Normal membership function, 228
Notice board, 168

Object, 166, 297
Object-oriented architecture, 297
Object-oriented implementation, 202
Object-oriented paradigm, 238
Object-oriented representation, 307
Open-loop control, 4
Optimal control, 15
    performance index, 15
Oscillations, 132
Overlap of membership, 95, 121
Overloaded component, 274, 276
Overshoot, 132, 217

Packaging workcell, 290
Paradoxes, 43
Parts-on-demand, 298
Peak load, 260

INDEX 341

Percentage overshoot, 116
Performance attribute, 132
Performance evaluator, 213, 218
Performance index, 308
Performance indicators, 213
Performance measure, 10
Performance monitoring, 253
Performance specifications, 213
Phase lead of compensator, 219
Phase margin, 219, 318
PID control, 5, 165, 211
PID servo, 152
Planner of restructuring, 287
Plausibility function, 98
Plausibility, 321
PopTalk language, 170
Possibility function, 317, 321
Power spectral density function, 15
PPI, see Proportional plus integral controller
Precision, 321
Predicate calculus, 29
Predicate, 289
Prefilter, 5
Premise, 28
Preprocessor, 82
Priority value, 277
Privileged match, 36
Probability function, 321
Probability, 67, 321
    subjective, 321
Procedural decomposition, 253
Process control, 252
Production level, 273
Production systems, 34
    forward (FPS), 35
Production workcell, 204
Projection, 55, 225
Prolog, 287, 290, 327
Proper subset, 49
Proportional gain, 6
Proportional plus integral controller (PPI), 155
Propositional calculus, 29
Propositions, 25

Qualitative terms, 236
Quality assessment, 265, 266
Quality control, 241
Quality index, 248
Quantization resolution, 73

Randomness, 322
Reasoning strategies, 35
Reasoning, 27
Reference model, 212, 217
Regulator, 4
Releasing of component, 278
Reliability, 277
Representation theorem, 50
Resolution function, 114
Resolution relationship, 112, 114
Resolution switching, 123
Resolution, 34
Restructuring problem, 272, 274
Restructuring, 269
Riccati equation, 16
Rise time, 217
Robotic applications, 129, 149
Robotic control system, 148
Robotic vision, 203
Robotics, 145
Robustness, 5, 14, 111, 318
    index, 318
Rule base tuning, 219
Rule dissociation, 112
Rule interaction, 99
Rule of inference, 28, 58
    compositional, 58
    conjunction (CRI), 28
    modus ponens (MPRI), 28
Rule validity, 324
Rule-base decoupling, 110
Rule-based systems, 34

S-norm, 66, 314
Schema, 167
Seam tracking experiments, 187
Secondary peak, 261
Self organization, 74, 165

Self-organization module, 305
Self-organizing rule base, 152
Self-tuning control, 10
Self-tuning controller, 232
Self-tuning system, 141
Semantic networks, 30
Sensing lines, 246
Sensor fusion, 305, 311
Separability of fuzzy restrictions, 227
Servo control, 5
Servo expert, 124, 158, 162, 166, 170, 213
Servo tuning, 216
Servo mechanism, 4
Set inclusion, 48
Shareability, 277
Sharing priority, 278
Sharing, 269
Signal combination, 251
Signal preprocessor, 5
Signum function, 13
Singleton function, 142
Singular value, 19
Sliding mode control, 12
        boundary layer, 14
        chattering, 14
        saturation function, 14
Sliding surface, 13
Smart products, 302
Soft (hard) algorithm, 105
Soft computing, 311
Soft control, 23
Specialized knowledge, 103, 265
Specifications, 171
Speed of response, 5, 116, 132
Stability of fuzzy systems, 115, 318
Stability region, 135
Stability, 5, 140
Stabilizability, 16
State equations, 15
Statistical process control (SPC), 39
Steady-state accuracy, 219
Steady-state error, 5, 116, 132
Structured information, 265

Subtask allocation, 253
Subtask modification, 253
Suction control, 12
sup (supremum) operation, 53, 72
Supervisory control, 147, 302, 305, 313
Support set, 79
Supremum (sup) operation, 313
Switching criteria, 12
Switching of fuzzy resolution, 122

T-conorm see S-norm
T-norm, 108, 314
Tactile sensing, 311
Task description, 253
Task sensor, 298
Task-sequence planner, 297
Template, 32
Threshold methods, 80
TIL software, see Togai
Togai Infralogic (TIL), 329
Top-down approach, 36
Toughest match, 36
Trajectory error, 131
Transfer function, 18
Transfer matrix, 18
Transitional operation, 157, 247, 307
Transitivity property, 65
Trapezoidal membership function, 93, 243
Triangular conorm, 314
Triangular membership function, 93, 243
Triangular norm, 108, 314
Truth table, 25
Tuner mapping module, 213
Tuner mapping relations, 220
Tuning action, 132
Tuning inference, 114
Tuning, 103

Uncertainty, 103, 321, 322
Uncoupled rule base, 226
Undercapacity component, 274, 276
Unified design, 309

Unimodal membership, 79
Union (OR), 25, 47
        generalized, 315
Universe of discourse, 44

Vagueness, 103, 321
Validity factor, 324
Variable structure control, 12
Venn diagram, 45
Vision, 311
Vision-based robotics, 203

White noise, 21
Workcell development, 253
Workcell, 204, 240, 269
Workload demand, 273
Workload of component, 273

Ziegler-Nichols method, 104